XI Workshop of the Gravitation and Mathematical Physics Division, Mexican Physical Society 2015

Journal of Physics: Conference Series Volume 1010

Guanajuato, Mexico
16 – 20 November 2015

ISBN: 978-1-5108-6287-6
ISSN: 1742-6588

Printed from e-media with permission by:

Curran Associates, Inc.
57 Morehouse Lane
Red Hook, NY 12571

Some format issues inherent in the e-media version may also appear in this print version.

Printed by Curran Associates, Inc. (2018)

For permission requests, please contact the Institute of Physics
at the address below.

Institute of Physics
Dirac House, Temple Back
Bristol BS1 6BE UK

Phone: 44 1 17 929 7481
Fax: 44 1 17 920 0979

techtracking@iop.org

Additional copies of this publication are available from:

Curran Associates, Inc.
57 Morehouse Lane
Red Hook, NY 12571 USA
Phone: 845-758-0400
Fax: 845-758-2633
Email: curran@proceedings.com
Web: www.proceedings.com

Table of contents

Volume 1010

XI Workshop of the Gravitation and Mathematical Physics Division of the Mexican Physical Society

16–20 November 2015, Guanajuato, Mexico

Accepted papers received: 16 April 2018
Published online: 2 May 2018

Preface

Papers

Abstract

The collection of papers in this volume was presented during the XI Workshop of the Gravitation and Mathematical Physics Division of the Mexican Physical Society (DGFM-SMF), which was held in Guanajuato, Guanajuato, México, November 16-20, 2015. The Workshop is a bi-annual series of conferences sponsored by the DGFM-SMF that started in 1993 with the purposes of discussing and exchanging the research and experience of the gravitational and mathematical physics communities in Mexico. Each Mexican Workshop has been devoted to subjects of broad interest, so that students, in particular, can have access to specialized courses and talks that allow them to raise up their qualifications as professional researchers. Recurrent topics in the Mexican Workshop are supergravity, branes, black holes, the early Universe, observational cosmology, quantum gravity and cosmology and numerical relativity.

This year, distinguished researchers in the field, working in Mexico, were invited to give courses, whereas young researchers were invited for plenary lectures. More specialized talks were also presented in parallel sessions, with ample participation of researchers, and graduate and undergraduate students. Unlike previous occasions the last day was devoted to celebrate the 100th anniversary of General Relativity.

The contributions in this volume have been peer-reviewed. We are indebted to the contributors of these proceedings, as well as to the other participants and organizers, all for making the event a complete success. We acknowledge the professionalism of our reviewers, who helped us to keep high quality standards in all manuscripts.

Acknowledgments

The organizing committee would like to acknowledge the financial support of the Mexican National Science and Technology Council (CONACyT), the Mexican Physical Society (SMF), as well as several Institutions including: Centro de Investigación y Estudios Avanzados (CINVESTAV), Universidad Autónoma Metropolitana Iztapalapa (UAM-I), Universidad de Guanajuato (UG), Universidad Nacional Autónoma de México (UNAM).

IOP Conf. Series: Journal of Physics: Conf. Series **1010** (2018) 011002 doi:10.1088/1742-6596/1010/1/011002

Peer review statement

All papers published in this volume of *Journal of Physics: Conference Series* have been peer reviewed through processes administered by the proceedings Editors. Reviews were conducted by expert referees to the professional and scientific standards expected of a proceedings journal published by IOP Publishing.

Canonical methods in classical and quantum gravity: An invitation to canonical LQG

Juan D Reyes
Centro de Ciencias Matemáticas,
Universidad Nacional Autónoma de México,
UNAM-Campus Morelia, A. Postal 61-3, Morelia, Michoacán 58090, Mexico

E-mail: jdreyes@matmor.unam.mx, jdrp75@gmail.com

Abstract. Loop Quantum Gravity (LQG) is a candidate quantum theory of gravity still under construction. LQG was originally conceived as a background independent canonical quantization of Einstein's general relativity theory. This contribution provides some physical motivations and an overview of some mathematical tools employed in canonical Loop Quantum Gravity. First, Hamiltonian classical methods are reviewed from a geometric perspective. Canonical Dirac quantization of general gauge systems is sketched next. The Hamiltonian formultation of gravity in geometric ADM and connection-triad variables is then presented to finally lay down the canonical loop quantization program. The presentation is geared toward advanced undergradute or graduate students in physics and/or non-specialists curious about LQG.

1. Why quantum gravity?

Classical general relativity is the most successful theory of gravity but it is incomplete. As made manifest by the celebrated singularity theorems due to Hawking and Penrose, on regimes of high energy density and strong curvature, geodesic incompleteness and other singularities signal the loss of predictability of the theory. On the other hand, quantum mechanics seems to describe remarkably well the physics down to the 10^{-19}m scale probed by particle accelerator experiments like the LHC. All other three known fundamental interactions in nature obey the rules of quantum mechanics, so we *believe* these rules should apply at a fundamental level, even at the Planck scale of 10^{-35}m!. In particular, we expect a fundamental description of the gravitational interaction to adhere to these rules and certainly, Einstein's theory does not.

A theory of gravity obeying quantum mechanics is expected to 'resolve' the classical singularities of general relativity (cosmological and black hole) along with other 'puzzles' of black holes (information loss, entropy count, etc.). It is also expected to serve as a natural regulator of ultravialotet divergences in QFT. Such sought for theory should do all this while at the same time recovering general relativity in the *semi-classical limit.*

The need for a quantum theory of gravity is not only conceptual. Even if quantization of gravity will not be directly detectable in the near future (e.g. F. Dyson), indirect evidence by its imprints on cosmological observations may be accessible in the not so distant future (e.g. D. Rideout et. al., G. Amelino-Camelia and L. Smolin, M. Bojowald and G. Calgani). Further constraints are also expected from gravitational wave observations.

In any case, one expects to learn a great deal and to get deep insights into the structure of spacetime by constructing a *consistent* quantum theory of gravity. There are however several

IOP Conf. Series: Journal of Physics: Conf. Series **1010** (2018) 012001 doi:10.1088/1742-6596/1010/1/012001

conceptual and technical problems one must tackle before claiming victory and it must be emphasized here that no such theory fulfilling the requirements or expectations above is available yet.

There are many (tentative) roads to quantum gravity. Amongst the most notable candidate theories or –more accurately– approaches to finding such theory are: String Theory, Non-commutative Geometry, Asymptotic Safety, Causal Sets, Causal Dynamical Triangulations (CDT) and Loop Quantum Gravity (LQG). Each path has its own limitations, advantages and perhaps modest but non trivial achievements.

1.1. Why Loop Quantum Gravity?

The path to quantum gravity one may take depends heavily on one's own prejudices about what physical (and correspondingly mathematical) principles must guide the construction of the theory and hold true in the quantum regime. This bias is usually tied to physicists' own training and skills. For example, high energy physicists may prefer String Theory or Asymptotic Safety scenarios, mathematically minded physicists may choose Non-commutativity, while relativists will perhaps lean towards background independent CDT or LQG. Whether any of these theories will lead to the 'correct' description of nature is –as we have said– still an open question.

In these notes we will consider LQG [1, 2, 3]. From the very beginning, the underlying philosophy behind the LQG program has been to take the lessons from Einstein's theory of gravity very seriously. The guiding concept in the construction of LQG is the physical principle of general relativity. This is mathematically encoded in the requirement of diffeomorphism invariance and background independence of the theory. The latter necessarily implies a *non-perturbative quantization* –as opposed to the perturbative approach of quantum field theory on Minkowski spacetime which has proved so successful in the description of the standard model of particle physics–. The universality of gravity makes it different from the other known interactions in nature and perhaps it should come as no surprise that different methods or description are needed.

There are several alternatives to constructing a LQG theory. These alternatives may be grouped in essentially two approaches:

- *Canonical (Dirac) quantization*
- Path integral-type quantizations (spin-foams)

Historically, a LQG theory was first obtained by applying well tested (Dirac) canonical quantization techniques to Einstein's general relativity theory. This procedure is –in the author's opinion– the more mathematically rigorous and solid way to construct kinematics. Path integral type quantizations were introduced later to include from the outset dynamics and explicit covariance in the framework. These two aspects have certainly proved harder to implement (or make explicit) in the canonical setting.

While the resulting (non separable) Hilbert spaces of LQG may seem 'exotic' compared to the good old Fock spaces used in particle physics, LQG is the result of applying well tested techniques and physical principles with only *minimal assumptions.* No extra dimensions or symmetries are required (although they may be incorporated in the framework) nor any supposition on non-commutativity. Under some mild assumptions, diffeomorphism invariance and background independence uniquely select the so called *loop representation.*

As we will review later, general relativity in Hamiltonian form may be expressed in terms of a so called connection and a triad. These canonical variables allow for natural smearings that enable one to construct a Poisson subalgebra of observables independent of any background structure or auxiliary metric: *the algebra of holonomies and fluxes.* The hallmark of loop quantization is to select and promote this particular set of observables and their Poisson bracket

IOP Conf. Series: Journal of Physics: Conf. Series **1010** (2018) 012001 doi:10.1088/1742-6596/1010/1/012001

relations to basic operators with corresponding commutation relations. From this basic set, all other operators in the quantum theory are to be constructed.

Gravity is geometry, and as a candidate quantum theory of gravity, LQG has shed light on several aspects of the quantum geometry of space, including a possible discreteness of space and some kinematical aspects of quantum black holes. Furthermore, the application of loop techniques in the quantization of symmetry-reduced mini or midisuperspace models has allowed the exploration of implications of quantum dynamics in these reduced settings and in particular, it gave birth to Loop Quantum Cosmology [4]. These results provide concrete mechanisms for singularity resolution, black hole entropy, and UV regulation for matter fields.

The kinematics of LQG are well understood. The kinematical Hilbert space and basic operators thereon can be rigorously constructed, there are no hidden infinities. Dynamics of the theory on the other hand, is not yet under control. Despite proposals for candidate Hamiltonian operators governing dynamics, or propagators in the corresponding manifestly covariant formulation of spin foams, there remain several technical and conceptual problems to be solved. In particular, for the canonical approach there are the quantization ambiguities in the definition of the Hamiltonian operator and most importantly, showing that these operators lead to a theory free of anomalies[1]. As we shall review briefly, this is a technical but crucial consistency condition, related to the covariance of the theory and to its semiclassical limit. Finding the latter is perhaps one of the most pressing issues also in the path integral formulation of spin foams. One may take Einstein's general relativity as a staring point for quantization, but a priori none of our known quantization recipes guarantees that the resulting quantum theory will reproduce –in an appropriately defined semiclassical regime– the classical theory one began with.

The problem of extracting dynamical information along with the semiclassical limit and physical predictions that could make LQG a falsifiable theory is compounded by the fact that semiclassical states are expected to be highly excited states, constructed nontrivially from the basic spin-network states of the kinematical Hilbert space. Even in symmetry reduced models, construction of Hamiltonian operators and the analysis of the dynamical difference equations they imply have proved difficult to deal with.

One possible route to overcome these problems and to arrive at physical implications of the quantum theory is to use effective classical equations [5, 6, 7], which are in closer contact with classical spacetime notions but that nevertheless incorporate corrections from a loop quantization. Indeed, effective equations have successfully been used in Loop Quantum Cosmology. For the simplest solvable systems they have been shown to closely follow quantum evolution not only for semiclassical regimes but even in the high energy density regime where the classical Big Bang singularity is replaced by a 'bounce'. Furthermore, by careful application of these effective techniques, phenomenology for the early universe that may be tested in future observations can be obtained [8]. We will however not delve into the canonical effective description here. For a brief introduction from a geometric perspective see e.g. [9].

The purpose of these notes is to make clearer some of the aspects and problems just exposed. We will not delve into the details of LQG, but rather focus on very basic notions and tools which are prerequisites to understanding these problems. Hopefully this will give the reader a general perspective on the subject and its methods, and serve as an incentive to get him or her more interested in this challenging but fascinating field.

For a more detailed but still concise introduction to LQG, the reader is referred to [1] and to [2, 3] for a more ample coverage.

[1] Not to mention other universal and long-standing conceptual issues like the problem of time.

2. Classical vs quantum descriptions

In physics, when we say we have a theory of some physical phenomena, what we imply is that we have a mathematical model characterizing and describing the behavior of some particular physical system or set of systems. This model includes some mathematical constructs which are supposed to completely codify all the relevant information one could possibly ask about the physical system and a correspondence between these mathematical objects and the things we can measure in a lab: the *observables.* While a purely *relational* description of the physical world including gravity is perhaps more fundamental, for physical phenomena more immediately accessible to our senses (and also for those far from our every day experience), it has been most fruitful or convenient to separate and formulate this description in terms *states* of the system and their *dynamical evolution.* For our everyday non relativistic theories –and even for relativistic ones– these states characterize the system 'at a given time' and their dynamical evolution describes their change 'with respect to time'.

Another ingredient in our formulations of physical reality are *symmetries* which play a major role for a concise description and four our understanding of physical theories. They are central in the gauge theories of the standard model and in general relativity.

One of the revolutions of the 20th century was the discovery that the behavior of physical systems at atomic and subatomic scales requires a mathematical description fundamentally different from the classical description tied to our every day experience. We believe this last found *quantum description* to be more fundamental, with the classical description being only an approximation. So given a classical theory like general relativity, how does one find the corresponding more fundamental quantum theory? Unfortunately, there is no straight forward answer to this question. Fortunately, there exist patterns and structures in both classical and quantum formulations which have given us general guidelines or *quantization procedures* to construct candidate quantum theories from classical ones. We are interested here in Dirac's canonical quantization procedure which –as we have said– can be followed to construct LQG. Dirac's procedure requires a classical theory to be in Hamiltonian form, so we will first go back to basics and review the Hamiltonian formulation putting emphasis on the relevant structures for quantization. We will compare the classical and quantum descriptions in terms of how the different aspects mentioned above: states, observables, dynamics and symmetries are implemented.

2.1. Classical description

Let us first recall how one arrives at a Hamiltonian formulation starting from a classical action. Consider for simplicity a non relativistic mechanical system with finite number N degrees of freedom q^i, generalized velocities $\dot{q}^i$, with $i = 1, \ldots, N$, and action

$$S = \int_{t_0}^{t^1} dt\, L(q^i, \dot{q}^i)\,.$$

For example, for n particles of mass m in $\mathbb{R}^3$, $N = 3n$, and our typical Lagrangian has the form

$$L = \frac{1}{2}\sum_{i=1}^{N} m\dot{q}^{i\,2} - V(q^i)\,,$$

for some potential function V. Variation of the action gives Euler-Lagrange equations of motion $\delta S = \frac{\partial L}{\partial q^i} - \frac{d}{dt}\frac{\partial L}{\partial \dot{q}^i} = 0$ resulting in this case in Newton's equations:

$$m\ddot{q}^i = -\frac{\partial V}{\partial q^i}\,.$$

To pass to the Hamiltonian formulation, one performs a Legendre transform:

$$q^i = q^i, \quad p_i = \frac{\partial L}{\partial \dot{q}^i}; \qquad H(q^i, p_i) = p_i \dot{q}^i - L\,.$$

The canonical action becomes

$$S = \int_{t_0}^{t^1} dt\, p_i \dot{q}^i - H(q^i, p_i)\,,$$

and its variation gives Hamilton's equations $\delta S = 0$:

$$\dot{q}^i = \frac{\partial H}{\partial p_i}\,, \qquad \dot{p}_i = -\frac{\partial H}{\partial q^i}\,. \tag{1}$$

The Legendre transform changes the second order differential equations of motion into a first order system for generalized positions q^i and momenta p_i. It follows that the state of the system at a given time t is completely determined by the values $\left(q^i(t), p_i(t)\right)$. All possible physical observables are simply functions of these $2N$ variables and their dynamics is governed by Hamilton's equations. What are then the underlying geometrical and algebraic structures behind this Hamiltonian formulation?

2.1.1. States: Classical phase space In the Hamiltonian formulation the state of the system is described by the generalized coordinates (q^i, p_i). The set of all possible states of a physical system defines its *phase space* Γ. This is a geometrical object: a manifold. For example, the state for a non relativistic point particle in three-dimensional space is determined by its position and momentum vectors in Cartesian coordinates: (X^i, P_i). These coordinate components may take arbitrary values, so the phase space is $\Gamma = \mathbb{R}^6$. In contrast, for a particle confined to move around a circle, phase space is a cylinder: $\Gamma = S^1 \times \mathbb{R}$, with coordinates given by the particle's angular position and momentum (θ, P_θ). The geometrical nature of phase space is clear if one recalls there is more than one way to describe a physical system and choose generalized coordinates for it. The state a physical system is in and the relations among different states are independent of these coordinates. This is essentially the defining property of a (topological) manifold. Since generalized coordinate transformations are further assumed to be smooth, Γ is a differentiable manifold. Each point on the manifold Γ represents a physical state of the system (Fig. 1).

The phase space of a classical theory is equipped with additional structures. Typical phase spaces obtained from a Lagrangian theory have the structure or may be identified with a special type of manifold[2]: the cotangent bundle T^*Q over the configuration space Q. This is a particular example of a more general class of manifolds called *symplectic*. Symplectic manifolds (Γ, Ω) are equipped with a special anti-symmetric bilinear form $\Omega_{\alpha\beta}$ (this is analogous to the physical four-dimensional spacetime manifold of special or general relativity which is equipped with a special symmetric bilinear form: the Minkowski metric $\eta_{\mu\nu}$ for special relativity or a more general pseudo-Riemannian metric $g_{\mu\nu}$ for general relativity). We will say more about this symplectic structure which plays a prominent role in the Hamiltonian description in the next subsections. All phase spaces we consider have a symplectic structure and are hence symplectic manifolds. Most important for canonical quantization is that on these spaces, one may define a *Poisson bracket* $\{\cdot,\cdot\} : C^\infty(\Gamma) \times C^\infty(\Gamma) \to C^\infty(\Gamma)$. This is a map, satisfying certain properties, that takes two differentiable functions on Γ and assigns another differentiable function on Γ. As it is standard, here $C^\infty(\Gamma)$ will denote the collection of all differentiable real functions on Γ.

[2] The Legendre transform defines a map between the tangent bundle manifold TQ, coordinatized by generalized positions and velocities: $(q^i, \dot{q}^i)$, and its dual cotangent bundle T^*Q, coordinatized by generalized positions and momenta.

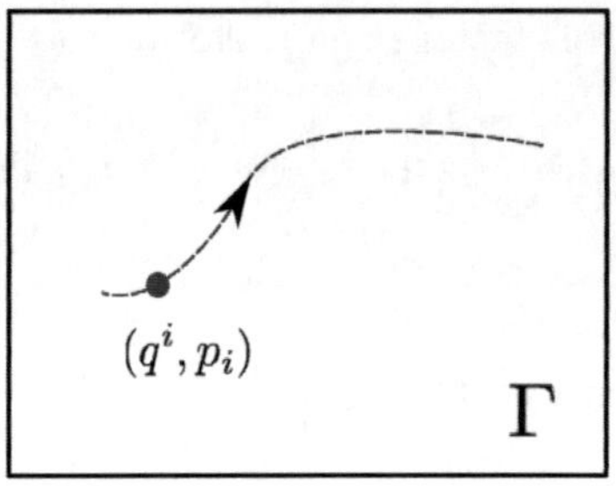

Figure 1. In the classical Hamiltonian formulation, a state is a point on phase space manifold Γ. As the system 'evolves', it describes a curve on phase space.

Going back to our original example of the phase space obtained from a Lagrangian theory, given the generalized coordinates (q^i, p_i) obtained after a Legendre transformation and for arbitrary functions $f, g \in C^\infty(\Gamma)$, the Poisson bracket is defined as

$$\{f, g\} = \sum_i \left(\frac{\partial f}{\partial q^i} \frac{\partial g}{\partial p_i} - \frac{\partial f}{\partial p_i} \frac{\partial g}{\partial q^i} \right). \tag{2}$$

However, we emphasize that just as points (and functions) on a manifold are independent of coordinates, this mapping of functions is also coordinate independent.

As a manifold, the phase space Γ may be coordinatized by different generalized or arbitrary coordinates $(\tilde{q}^i, \tilde{p}_i)$. For a general phase space, *canonical* or *Darboux* coordinates (q^i, p_i) are precisely defined to be coordinates for which the Poisson bracket is expresssed as in (2). For canonical coordinates, we have the *Canonical Relations*:

$$\{q^i, p_j\} = \delta^i_j, \qquad \text{and} \qquad \{q^i, q^j\} = \{p_i, p_j\} = 0\,.$$

Just for fun, we can abstract things even further. Symplectic manifolds are actually a subclass of a more general type of manifolds: *Poisson manifolds* $(\Gamma, \{\cdot,\cdot\})$. These are manifolds Γ equipped with a bracket $\{\cdot,\cdot\} : C^\infty(\Gamma) \times C^\infty(\Gamma) \to C^\infty(\Gamma)$ acting on functions on the manifold and satisfying the basic properties that (2) does[3]. Every phase space being a symplectic manifold is a Poisson manifold and its bracket may be written as in (2), but the converse is not true. Not every Poisson manifold is symplectic and its bracket may not have the form (2).

So far we have focused on mechanical systems with only a finite number of degrees of freedom for which phase spaces are finite dimensional. For field theories like Maxwell's electromagnetism (or gravity) there is an infinite number of degrees of freedom. Essentially, for each point $\mathbf{x} = (x^a)$ on three-dimensional space, there is a set of canonical pairs. For example, for electromagnetism the pair defined by the components of magnetic potential and electric field $\left(\vec{A}(\mathbf{x}), \vec{E}(\mathbf{x})\right)$ defines canonical variables at each point of space. The corresponding phase spaces are therefore infinite dimensional manifolds. Nevertheless –at least formally– all the constructions above generalize to this case[4].

[3] The bracket is antisymmetric: $\{f, g\} = -\{g, f\}$, bilinear: $\{f, g + h\} = \{f, g\} + \{f, h\}$, satisfies the Jacobi identity: $f\{g, h\} + g\{h, f\} + h\{f, g\} = 0$, and the Leibniz rule: $\{f, gh\} = g\{f, h\} + \{f, g\}h$.

[4] Infinite dimensional manifolds pose many technical complications. Points on phase space are identified by functions on physical space rather than a finite set of numbers. A complete description of phase space requires then a specification of the type of functions considered. On these functional spaces there are –among other things– convergence issues which need to be considered carefully in order to properly define differentiable structures thereon. There is an alternative description of the formalism –perhaps more elegant but less intuitive than infinite dimensional manifolds– in terms of so called *jets*, but it too comes with its own technical nuisances.

IOP Conf. Series: Journal of Physics: Conf. Series **1010** (2018) 012001 doi:10.1088/1742-6596/1010/1/012001

2.1.2. Observables: Poisson algebras For theories without redundancies in their mathematical description or so called local gauge symmetries, the correspondence between the mathematical constructs defining the classical theory and the physical observables one measures in a laboratory is in principle straight forward. Observables in the theory correspond to $C^\infty(\Gamma)$, the set of all differentiable functions f on Γ:

$$f := \Gamma \to \mathbb{R}$$

Prime examples are the total energy $H(q,p)$ in a conservative system or the very same generalized coordinate functions $\pi^i(q^j,p_j) = q^i$, $\pi_i(q^j,p_j) = p_j$ of non relativistic mechanical theories, which may represent coordinates and momenta of individual components or particles of the system.

At the end of the day, physical observables are the most important, or even the only thing, one cares about. It is not surprising then that relations among these functions play a central role in the description of the classical theory and its quantization. These relations are algebraic in nature. Even mathematically, the underlying geometry of phase space may be subordinated to these algebraic relations and reconstructed from them. This is key in canonical quantization and it is also at the root of approaches to quantization of gravity like noncommutative geometry. Intuitively, one may guess this encoding to be true since the coordinate functions mapping and differentiating points on phase space are a part of the set of observable functions.

What are these referred algebraic structures? Let us first recall the mathematical definition of an *algebra*. An algebra $\mathfrak{A}$ is a vector space (over $\mathbb{R}$ or $\mathbb{C}$) equipped with a multiplication of vectors compatible with the sum and scalar product, what this means is that for any vectors $A, B, C \in \mathfrak{A}$ and scalars $a, b, c \in \mathbb{C}$, this multiplication satisfies expected properties:

$$A(bB + cC) = bAB + cAC \qquad \text{and} \qquad (aA + bB)C = aAC + bBC\,.$$

An algebra is said to be *associative* if the product further satisfies: $A(BC) = (AB)C$. It is *abelian* if the product of two vectors commutes: $AB = BA$. If the product of an algebra is neither abelian nor associative but instead it is anti-symmetric and satisfies the *Jacobi identity*, i.e.

$$AB = -BA \qquad \text{and} \qquad A(BC) + B(CA) + C(AB) = 0\,,$$

then $\mathfrak{A}$ is called a *Lie algebra*[5].

The set of observables $C^\infty(\Gamma)$ is naturally an abelian associative algebra with product defined pointwise, but most importantly, it is also a Lie algebra with respect to a product defined by the Poisson bracket. Indeed, $C^\infty(\Gamma)$ is clearly a vector space with sum and product by scalar c defined pointwise:

$$(f+g)(x) := f(x) + g(x), \qquad (cf)(x) := cf(x)\,, \qquad \forall x \in \Gamma\,.$$

$C^\infty(\Gamma)$ can be made an abelian associative algebra with product also defined pointwise:

$$(fg)(x) := f(x)g(x). \tag{3}$$

However, since phase space Γ is a Poisson manifold, the Poisson bracket $\{\cdot,\cdot\} : C^\infty(\Gamma) \times C^\infty(\Gamma) \to C^\infty(\Gamma)$ defines another product:

$$f \cdot g := \{f, g\}.$$

It is trivial to see that the properties of the Poisson bracket are equivalent to this product being compatible with the linear structure of $C^\infty(\Gamma)$ and satisfying

$$f \cdot g = -g \cdot f \qquad \text{and} \qquad f \cdot (g \cdot h) + g \cdot (h \cdot f) + h \cdot (f \cdot g) = 0\,.$$

[5] Do not be confused. The (finite dimensional) Lie algebras associated with good old Lie groups describing symmetries in physics are but one class in this more general definition.

Thus, the set of observable functions $C^\infty(\Gamma)$ is a Lie algebra with the product defined by the Poisson bracket. Notice however that, even for finite dimensional phase spaces, the algebra is infinite dimensional (as a vector space). This algebra is what one calls the *Poisson algebra of observables*, and it is the relevant algebraic structure for quantization.

For theories with local gauge symmetries, there is a complication. Due to the ambiguities in the description of the system, which translate to different points on Γ representing the same physical state, $C^\infty(\Gamma)$ does not directly correspond to actual physical observables we measure. In order to set this correspondence, one needs to further 'factor out' these ambiguities. We will comment more on this later.

2.1.3. Dynamics: Symplectic geometry Evolution is given by Hamilton's equations

$$\dot{q}^i = \frac{\partial H}{\partial p_i} = \{q^i, H\}\,, \qquad \dot{p}_i = -\frac{\partial H}{\partial q^i} = \{p_i, H\}\,,$$

or for a general function $f(q,p)$ on phase space:

$$\dot{f} = \{f, H\}\,.$$

But how does one describe dynamical evolution in the geometric picture given by phase space? or in other words, what do Hamilton equations tell us about the geometry of phase space and viceversa? Hamiltonian evolution is characterized by two structures: a privileged real function on Γ, i.e. the Hamiltonian H (usually associated with the energy of the system, e.g. $H = p_i^2/2m + V(q^i)$ in mechanics), and the symplectic structure Ω.

To see how such structures emerge from Hamilton's equations (1), we may first write these equations in block matrix form:

$$\begin{pmatrix} \dot{q}^i \\ \dot{p}_i \end{pmatrix} = \begin{pmatrix} 0 & \mathbb{I} \\ -\mathbb{I} & 0 \end{pmatrix} \begin{pmatrix} \frac{\partial H}{\partial q^i} \\ \frac{\partial H}{\partial p_i} \end{pmatrix}, \tag{4}$$

with $\mathbb{I}$ denoting the $N \times N$ identity matrix. Let us now recall that canonical transformations are also defined as those transformations leaving Hamilton's equations unchanged. One can then see that

$$\mathcal{P} := \begin{pmatrix} 0 & \mathbb{I} \\ -\mathbb{I} & 0 \end{pmatrix}, \qquad \text{and therefore also its inverse:} \quad \Omega := \begin{pmatrix} 0 & \mathbb{I} \\ -\mathbb{I} & 0 \end{pmatrix},$$

are invariant under canonical transformations or equivalently, they are independent of canonical coordinates. But there is more. At each point on phase space, Ω defines a nondegenerate, anti-symmetric bilinear transformation of tangent vectors. In the language of differential geometry, Ω is a nondegenerate two-form. This is the *symplectic structure* we referred to before. It is a geometrical object, independent of any coordinates on Γ.

The symplectic structure defines an isomorphism between tangent vectors and their dual one-forms. In other words, we can use $\Omega_{\alpha\beta}$ to 'lower' vector indices: $X_\alpha = \Omega_{\alpha\beta}X^\beta$, and we can use its inverse $\Omega^{\alpha\beta} := \mathcal{P}^{\alpha\beta}$ to 'raise' these indices back: $X^\alpha = \Omega^{\alpha\beta}X_\beta$ (just like one uses the metric $g_{\mu\nu}$ and its inverse $g^{\mu\nu}$ to 'lower' and 'raise' indices of tensors in special or general relativity). For our examples here, indices (matrix components) run as $\alpha, \beta = 1, \cdots, 2N$.

The inverse of the symplectic structure $\mathcal{P} = \Omega^{-1}$ is also called the *Poisson bivector* and there is a reason for that. The Poisson bracket is really this bivector acting on differentials

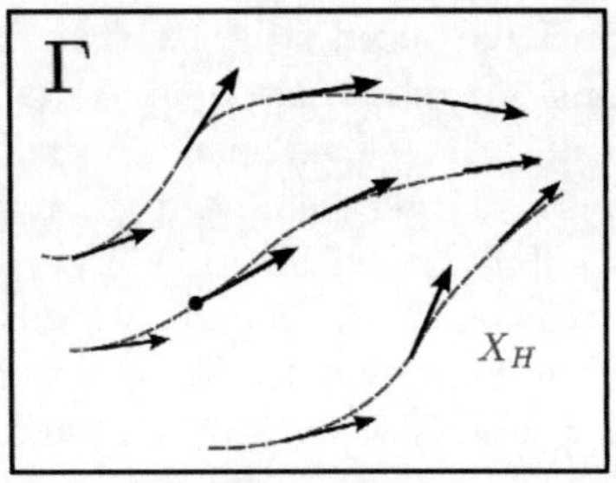

Figure 2. Hamiltonian evolution on phase space moves along the integral curves of the Hamiltonian vector field X_H determined by the physical Hamiltonian function H.

$\mathrm{d}f = \frac{\partial f}{\partial q^i}\,\mathrm{d}q^i + \frac{\partial f}{\partial p_i}\,\mathrm{d}p_i$ of functions f[6]:

$$\{f, g\} = \sum_i \frac{\partial f}{\partial q^i}\frac{\partial g}{\partial p_i} - \frac{\partial f}{\partial p_i}\frac{\partial g}{\partial q^i} = \mathcal{P}^{\alpha\beta}(\mathrm{d}f)_\alpha(\mathrm{d}g)_\beta\,. \tag{5}$$

To complete the geometric picture of Hamiltonian evolution, we observe that the symplectic structure (or equivalently the Poisson bracket) provides a very neat way to construct vector fields on Γ. Indeed, given a function f, its differential $\mathrm{d}f$ determines a vector field X_f thanks to the isomorphism –defined by the symplectic structure– between tangent vectors and their dual one-forms. [From (5), this vector field is sometimes also denoted as $X_H = \{\cdot, H\}$]. So given the Hamiltonian function of the system $H : \Gamma \to \mathbb{R}$, by raising the indices of its differential $\mathrm{d}H$, we get a *Hamiltonian vector field*[7]:

$$X_H^\alpha := \mathcal{P}^{\alpha\beta}(\mathrm{d}H)_\beta\,.$$

Now the picture is complete. As the physical system evolves in time, it describes a curve $\gamma(t) := \big(q^i(t), p_i(t)\big)$ on phase space (Fig. 1). What Hamilton's equations (4) tell us is that this evolution is along directions determined by the Hamiltonian vector field X_H (Fig. 2). Indeed, (4) may be rewritten as

$$\dot{\gamma}(t) = X_H\,,$$

which says $\gamma(t)$ is an integral curve of X_H (a curve whose tangent velocities $\dot{\gamma}(t)$ coincide with the vector field X_H).

2.1.4. Symmetries: Gauge theories and constrained systems Symmetries are very important in physics. Symmetries may be global (acting on the physical system as a whole) or local (acting independently on individual parts of the system, e.g. on fields at each point in space). Here we will focus on the latter. All our fundamental theories of nature: the standard model and general relativity, are *gauge theories*. This is to say they are theories with local symmetries. Intuitively, local symmetries reflect our freedom to change the 'reference frame' with respect to which we describe the physical system locally. So different (mathematical) states will result from changing to different 'reference frames' locally, but they will all represent or correspond to the same physical state.

Local symmetries are hence a redundancy in the mathematical description of the theory. This redundancy translates into two facts in the Hamiltonian description:

[6] The Jacobi identity satisfied by the Poisson bracket implies another defining property of the symplectic structure: its differential vanishes $\mathrm{d}\Omega = 0$, i.e. Ω is a closed.

[7] To confuse the uninitiated, any vector field X_f obtained this way from an arbitrary function f is also called Hamiltonian.

(i) The number of (canonical) variables used is larger than what is really needed to describe the system at a given 'instant' of its evolution, but eliminating the redundant variables is not trivial. More specifically, the phase space Γ that results from allowing arbitrary values of the canonical variables is larger than needed. Actual physical states and evolution are necessarily confined to a subregion within this larger space Γ. This subregion may be characterized by a set of equations for the canonical variables (just like the unit circle on the (x, y)-plane is characterized by the equation $x^2 + y^2 - 1 = 0$). In other words, for gauge theories there are non trivial relations or *constraints* among the canonical variables:

$$C^I(q^i, p_i) \approx 0 \,. \tag{6}$$

These constraint equations define the *constraint (hyper) surface* on Γ, to which physical states and evolution are restricted (Fig. 3). The index I runs over the number of constraints necessary to characterize this surface.

(ii) The redundancy also implies that different points on Γ may represent the same physical state. These points are necessarily related by a local (gauge) symmetry. Starting from one point on the constraint surface and applying successively (infinitesimal) gauge symmetry transformations, one may obtain all points in Γ representing the same physical state as the original point. These are the *gauge orbits* (Fig. 3).

When one 'changes the gauge', i.e. when one applies a local symmetry transformation, the physical system does not change but the actual point on Γ representing the system does. This transformation should not change the (Hamiltonian) equations of motion, so mathematically it is analogous to dynamical evolution. (Infinitesimal) gauge transformations are therefore also represented by Hamiltonian flows, i.e. they correspond to Hamiltonian vector fields X_{C^I} defined using generator functions $C^I : \Gamma \to \mathbb{R}$ as in the previous discussion on dynamics:

$$\delta q^i = \epsilon\{q^i, C^I\}, \qquad \delta p^i = \epsilon\{p^i, C^I\}, \qquad \epsilon << 1 \,.$$

The same geometric picture for evolution (Fig. 2), with X_H replaced by X_{C^I}, is valid for gauge transformations.

The notation for the generator functions of gauge transformations is no coincidence. Generically, if the constraint functions in (6) are –in Dirac's terminology– *first class*, i.e. their Poisson bracket is a linear combination of constraints:

$$\{C^I, C^J\} = f^{IJ}_K C^K \,, \tag{7}$$

for some functions $f^{IJ}_K(q, p)$, then one can show they also correspond to generator functions for vector fields X_{C^I} generating gauge symmetry orbits.

For an in depth discussion of the Hamiltonian description and constrained systems, the reader is referred to the excelent references [10, 11, 12].

2.2. Quantum description

We now turn to quantum mechanics. At the center of the mathematical quantum description, there is the notion of Hilbert spaces and operators thereon. The *probabilistic interpretation* then connects these mathematical objects to physical or measurable observables. While there are important generalizations or reformulations of the quantum theory which do not rely directly on these mathematical notions, for the moment we will not consider such extensions, and focus on the Hilbert space formulation canonical quantization takes aim at. Again, we will briefly recall how states, observables, dynamics and symmetries of the physical system are represented in this case.

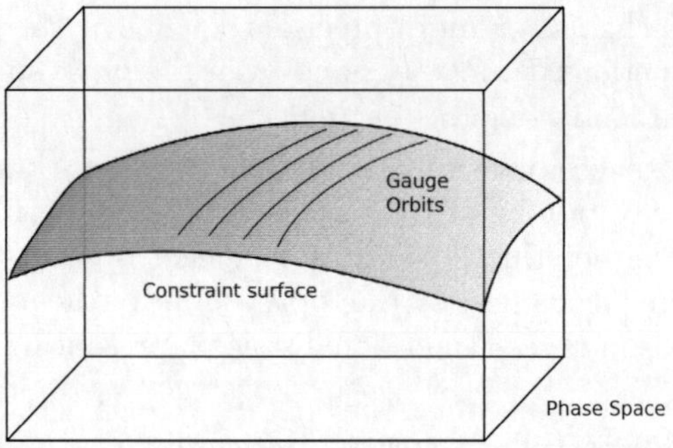

Figure 3. Geometry of constraints: Evolution and states are restricted to the constraint surface on Γ and different points representing the same physical state define gauge orbits. First class constraint functions play a dual role in the Hamiltonian formulation, they define the constraint surface and they are the canonical generators of gauge transformations.

2.2.1. States: Hilbert spaces In the quantum description, the states of a physical system are represented by normalized vectors $|\Psi\rangle$ (or more accurately by rays $e^{i\theta}|\Psi\rangle$) in a Hilbert space $\mathcal{H}$.

Hilbert spaces generalize notions on finite dimensional Euclidean vector spaces –with their 'dot product'– to include infinite dimensional vector spaces. More precisely, a Hilbert space $\mathcal{H}$ is a complex vector space equipped with an *inner product* $\langle\cdot|\cdot\rangle : \mathcal{H}\times\mathcal{H}\to\mathbb{C}$ satisfying:

(i) $\langle\Phi|\alpha\Psi_1+\beta\Psi_2\rangle = \alpha\langle\Phi|\Psi_1\rangle+\beta\langle\Phi|\Psi_2\rangle$, para $\Phi, \Psi_1, \Psi_2 \in \mathcal{H}$ y $\alpha, \beta \in \mathbb{C}$.

(ii) $\langle\Phi|\Psi\rangle = \overline{\langle\Psi|\Phi\rangle}$ (bar denoting complex conjugation).

(iii) $\langle\Phi|\Phi\rangle \geq 0$, with equality iff $\Phi = 0$.

The inner product induces a norm $||\Phi|| := \sqrt{\langle\Phi|\Phi\rangle}$ and therefore a metric $d(\Phi,\Psi) := ||\Phi-\Psi||$ on $\mathcal{H}$. The metric introduces topological notions on $\mathcal{H}$, in particular the notion of convergence of sequences of vectors. As a metric space, $\mathcal{H}$ is required to be *complete* (a complete space contains limit points of all its Cauchy sequences).

The completeness condition guarantees that on every Hilbert space there exist an *orthonormal basis*. This is a set of vectors $\{|e_\alpha\rangle\}$ in $\mathcal{H}$ satisfying

$$\langle e_\alpha|e_\beta\rangle = \delta_{\alpha,\beta}$$

and such that for any vector $|\Psi\rangle \in \mathcal{H}$ we have an expansion of the form

$$|\Psi\rangle = \sum_n \langle e_{\alpha_n}|\Psi\rangle|e_{\alpha_n}\rangle\,, \qquad n = 1, 2, \ldots$$

We want to stress that for each vector $|\Psi\rangle$, the number of basis elements in the previous sum may be finite or infinite, but always numerable ($n \in \mathbb{N}$). However, notice that we have not specified the index set for the labels α identifying each element $|e_\alpha\rangle$ of the basis. This is the number of elements in the basis necessary to expand every vector in $\mathcal{H}$ in this way. A Hilbert space may be classified by the cardinality of its orthonormal basis:

- *Separable* Hilbert spaces have a countable basis, that is either
  - a finite basis $\{e_n\rangle\}_{n=1,\ldots,M}$, like the Cartesian product $\mathbb{C}^M$
  - or a countable infinite basis $\{e_n\rangle\}_{n\in\mathbb{N}}$. Typical examples are l^2, the set of square summable $\sum_{n=1}^{\infty}|z_n|^2 < \infty$ infinite sequences $(z_1, z_2, \ldots)$, the 'wave functions' of

basic quantum mechanis, or more precisely $L^2(\mathbb{R}, \mathrm{d}x)$ the space of square integrable $\int_{\mathbb{R}} \mathrm{d}x|f|^2 < \infty$ functions $f : \mathbb{R} \to \mathbb{C}$, and even the Fock spaces used in quantum field theory they are all Hilbert spaces of this type.

- *Non-separable* Hilbert spaces have an uncountable basis $\{e_\alpha\rangle\}_{\alpha \in \mathbb{R}}, \ldots$. That is, bases with the cardinality of the real numbers $\mathbb{R}$ (or bigger!). The Hilbert spaces resulting from loop quantization are precisely of this type. Bases labeled by points on the real line as in so called polymer quantum mechanics, by the total volume of the universe as in Loop Quantum cosmology, or by graphs in three-dimensional space representing excitations of geometry as in full LQG.

Actually, in the category of Hilbert spaces, the cardinality or number of elements in an orthonormal basis completely characterizes the Hilbert space. If two Hilbert spaces $\mathcal{H}$ and $\tilde{\mathcal{H}}$ have corresponding bases $\{|e_\alpha\rangle\}_{\alpha \in I}$ and $\{|\tilde{e}_\alpha\rangle\}_{\alpha \in I}$ labeled by the same index I, the mapping $U : \mathcal{H} \to \tilde{\mathcal{H}}$, $|e_\alpha\rangle \mapsto |\tilde{e}_\alpha\rangle$, for each $\alpha \in I$ defines an isometry. In particular, all infinite dimensional separable Hilbert spaces are isometric to the seemingly more simple l^2 space. Nevertheless, as models of physical systems, it may be certainly more convenient to work with or construct separable Hilbert spaces which do not resemble l^2. When Heisenberg introduced his 'matrix quantum mechanics' he was essentially working with l^2. At the same time Schrödinger, with his 'wave functions' for particles in one dimension, was using $L^2(\mathbb{R}, \mathrm{d}x)$. The latter realization of the Hilbert space for the quantum behavior of a point particle is equivalent to l^2, but it has the virtue of having a more direct connection or physical interpretation in terms of positions, momenta and undulatory phenomena associated with the way we measure quantum properties of particles in a laboratory.

These observations already make manifest a difference between the role of states in classical and quantum (mathematical) descriptions. Whereas in the classical formulation, all physical information is essentially already encoded in the states, i.e. in the manifold Γ (which mathematically –using the symplectic structure– determines the algebra of observables), in the quantum description, the Hilbert space $\mathcal{H}$ by itself does not contain or codify the whole information about the physical system. Additional (external) input is required to complete the mathematical description. This additional information is codified in the quantum algebra of observables.

2.2.2. Observables: operator algebras The correspondence between the mathematical objects defining the quantum theory and the physical quantities one may measure in a lab is less direct than in the classical theory. Each physical observable is associated in the quantum theory with a linear operator on the Hilbert space:

$$\widehat{f} : \mathcal{H} \to \mathcal{H} .$$

The value λ, a physical observable may possibly take after a measurement is necessarily an eigenvalue of its corresponding operator

$$\widehat{f}|\Psi\rangle = \lambda|\Psi\rangle .$$

Real observables then generally correspond to *Hermitian* or *self-adjoint* operators

$$\widehat{f} = \widehat{f}^\dagger ,$$

which ensures their eigenvalues are real.

The inner product and normalization of vectors make possible the *probabilistic interpretation* which provides physical predictions through expectation values of measured observables on a state $|\Psi\rangle$ and transition probability amplitudes between states:

$$\langle\widehat{f}\rangle = \langle\Psi|\widehat{f}|\Psi\rangle \qquad \text{and} \qquad P(|\Psi\rangle \to |\Phi\rangle) = |\langle|\Phi|\Psi\rangle|^2 .$$

Just like the algebra of observable functions on phase space in a classical description, the set of *observable operators* (operators associated with physical observables) forms a (Lie) algebra. Indeed, observable operators are a subset of $GL(\mathcal{H})$, the set of general linear operators on $\mathcal{H}$. $GL(\mathcal{H})$ not only has the structure of a vector space, but also of an associative algebra and of a Lie algebra. Vector addition of operators $\widehat{f}+\widehat{g}$ and scalar multiplication $c\widehat{f}$ are defined by its action on vectors $|\Psi\rangle \in \mathcal{H}$ as

$$(\widehat{f}+\widehat{g})|\Psi\rangle := \widehat{f}|\Psi\rangle + \widehat{g}|\Psi\rangle\,, \qquad (c\widehat{f})|\Psi\rangle := c\widehat{f}|\Psi\rangle\,.$$

The product given by the composition $\circ$ of operators:

$$\widehat{f}\,\widehat{g} := \widehat{f}\circ\widehat{g}\,,$$

makes $GL(\mathcal{H})$ a noncommutative associative algebra [in contrast to the abelian associative algebra defined by classical observable functions with pointwise product (3)]. Using this composition product, the *commutator*:

$$[\widehat{f},\widehat{g}] := \widehat{f}\,\widehat{g} - \widehat{g}\,\widehat{f}\,,$$

(defines yet another product which) makes $GL(\mathcal{H})$ a Lie algebra. The set of observable operators in necessarily a subalgebra of it.

Analogously to its classical counterpart, due to the chief role of physical observables, the latter algebra of operator observables is a key ingredient of the quantum description. The 'abstract algebra' is however not enough. To extract physical information one further needs a *state.* This is where the *representation*, i.e. the actual Hilbert space with its inner product and specific operators, plays its part.

2.2.3. Dynamics Dynamical or time evolution must preserve the probabilistic interpretation and the *superposition principle*, i.e. it must preserve the normalization of vectors and it must be linear. Evolution must therefore be implemented by unitary transformations $\mathbf{U}_t : \mathcal{H} \to \mathcal{H}$:

$$|\Psi(t+t_0)\rangle = \mathbf{U}_t|\Psi(t_0)\rangle\,.$$

For simple (e.g. linear) unconstrained systems, this evolution operator may be (infinitesimally) generated by a *Hamiltonian operator* $\widehat{H}$ (see next subsection). In this case, evolution of a state is dictated by the *Schrödinger equation*:

$$i\hbar\frac{\partial}{\partial t}|\Psi\rangle = \widehat{H}|\Psi\rangle\,.$$

Since the information about the physical system is honestly split mathematically between the states and the algebra of observable operators in the quantum theory, it is not surprising that evolution may be equivalently thought as evolution of the observable operators (Heisenberg picture):

$$\widehat{f}(t+t_0) = \mathbf{U}_{-t}\,\widehat{f}(t_0)\,\mathbf{U}_t\,,$$

as opposed to evolution of the states (Schrödinger picture). Schrödinger equation is then equivalent to *Heisenberg equation*

$$i\hbar\frac{\partial\widehat{f}}{\partial t} = [\widehat{f},\widehat{H}]\,.$$

Notice the special role of t (and the Hamiltonian $\widehat{H}$) in these quantum descriptions. Schrödinger or Heisenberg equations do not easily generalize to more complicated systems. This is specially true for totally constrained systems –which have vanishing Hamiltonian– like general relativity.

IOP Conf. Series: Journal of Physics: Conf. Series **1010** (2018) 012001 doi:10.1088/1742-6596/1010/1/012001

2.2.4. Symmetries Symmetries (global and local) should not change physical predictions of the quantum theory and therefore are also *represented* by unitary transformations $\mathbf{U}_g : \mathcal{H} \to \mathcal{H}$ (one for each symmetry transformation g) acting on states $|\Psi\rangle$ and observable operators $\widehat{f}$:

$$|\Psi\rangle \to \mathbf{U}_g|\Psi\rangle\,, \qquad \widehat{f} \to \mathbf{U}_g\,\widehat{f}\,\mathbf{U}_g^{-1}\,.$$

Recall, an operator $\mathbf{U}$ is *unitary* iff

$$\mathbf{U}\mathbf{U}^\dagger = \mathbb{I}, \quad \text{or equivalently } \mathbf{U}^{-1} = \mathbf{U}^\dagger\,,$$

so indeed physical predictions –eigenvalues, expectation values, transitions amplitudes– are unchanged under such transformations (e.g. for the transition amplitude between two states: $\langle \mathbf{U}\Phi|\mathbf{U}\Psi\rangle = \langle\Phi|\mathbf{U}^\dagger\mathbf{U}|\Psi\rangle = \langle\Phi|\Psi\rangle$).

The family $\{\mathbf{U}_g\}_{g\in G}$ gives a unitary *representation* of the symmetry group of transformations G, this means it satisfies

$$\mathbf{U}_{g\circ h} = \mathbf{U}_g\mathbf{U}_h, \qquad \mathbf{U}_{g^{-1}} = \mathbf{U}_g^{-1}\,, \qquad \forall\, g,h \in G\,,$$

where $\circ$ denotes the composition of transformations (the product of the group) and g^{-1} the inverse transformation of g. The symmetry group G may consist of global transformations and consequently be relatively 'small' (like the compact Lie group of global rotations in Euclidean 3-space $SO(3)$ or the non-compact Poincaré group of global symmetries in Minkowski 4-space $\mathcal{M}$) but it may also be a 'bigger' gauge group (like the gauge group $\mathcal{G}$ of the standard model or the group of diffeomorphisms of spacetime Diff($\mathcal{M}$) in general relativity[8]).

If $g(\beta)$ with $\beta \in \mathbb{R}$ is a one-parameter family of symmetry transformations, that is the family of transformations labeled by β satisfies $g(\beta_1 + \beta_2) = g(\beta_1) \circ g(\beta_2)$, at least formally, we can build transformation $\mathbf{U}_{g(\beta)}$ by lots of little ones:

$$g(\beta) = \underbrace{g(\beta/N) \circ \cdots \circ g(\beta/N)}_{N\text{factors}}\,,$$

so

$$\mathbf{U}_{g(\beta)} = [\mathbf{U}_{g(\beta/N)}]^N\,.$$

In the limit $N \to \infty$, these pieces are infinitesimal transformations which are 'almost the identity transformation' and therefore may be written as:

$$\mathbf{U}_{g(\epsilon)} = \mathbb{I} + i\epsilon\widehat{C}\,, \qquad \epsilon << 1$$

A finite transformation is then recovered from $\widehat{C}$:

$$\begin{aligned}\mathbf{U}_{g(\beta)} = \lim_{N\to\infty}\left[\mathbb{I} + i(\beta/N)\widehat{C}\right]^N &= \exp(i\beta\widehat{C}) \\ &= \mathbb{I} + i\beta\widehat{C} + \frac{1}{2}(i\beta\widehat{C})^2 + \cdots\end{aligned}$$

$\widehat{C}$ is the *infinitesimal generator* of the one-parameter family of transformations.

A state $|\Psi\rangle$ representing a configuration of the physical system invariant under symmetry transformations must satify $\mathbf{U}_{g(\beta)}|\Psi\rangle = |\Psi\rangle$, and hence it also satisfies $\widehat{C}|\Psi\rangle = 0$.

[8] The elements $g \in \mathcal{G}$ of the gauge group of the standard model are functions on Minkowski spacetime $g : \mathcal{M} \to SU(3) \times SU(2) \times U(1)$ and the elements of Diff($\mathcal{M}$) are diffeomorphisms $g : \mathcal{M} \to \mathcal{M}$, or equivalently coordinate transformations of the spacetime manifold $\mathcal{M}$.

3. Canonical quantization

The behavior of atomic and subatomic systems requires quantum mechanics, so we believe quantum theory to be more fundamental (and richer) than the classical description which is supposed to emerge from the first as an effective or coarse-grained description. Unfortunately, we have no physical first principles for quantum theory yet. To construct a quantum theory for a physical system from scratch, we really only have the *correspondence principle* as guidance: Excluding certain quantum observables –like the spin of particles– which have no classical analog, the quantum theory is expected and must be made to correspond to the classical mechanics (or field theory) one started with in an appropriately defined *semi-classical regime.* That is, through the *semi-classical limit*, where by some kind of generalization of Erhenfest theorem, expectation values of quantum operators in so called *semi-classical states* should reproduce the classical dynamics.

Of course in principle, to construct a quantum theory of a given physical system, one could devise and perform a great many experiments on the system (perhaps infinitely many!) and cleverly organize and synthesize the results according to the basic (mathematical) rules of quantum mechanics[9]. Regardless of being unsatisfactory from a theoretical point of view, this way of proceeding is generally impractical, and for systems like gravity, technologically impossible today (and must likely for many more decades to come). The only known systematic way to formulate a possible quantum theory describing a physical system is through *quantization* of the corresponding classical description. This is the process of building the seemingly very different mathematical objects describing states, observables, evolution and symmetries of a candidate quantum theory from their mathematical counter parts in the classical theory, in such a way that the correspondence principle is satisfied.

So given a classical theory, how do we actually quantize it? how do we find the particular Hilbert space $\mathcal{H}$, operators $\widehat{f}$ and all that which will actually describe the quantum system? There are several general approaches or rather *recipes* to finding such theory. To name the 'most popular' we have

- Canonical quantization
- Path integral quantization
- Geometric quantization
- Deformation quantization

with the first two probably being at the top of physicists' likings. While observables are the heart of both classical and quantum descriptions, each procedure privileges certain objects over the others and uses them differently to construct a candidate theory.

Notice we have said quantization of a classical theory in general, and in particular the procedures mentioned above, result not in *the* quantum theory, but in a *candidate* quantum theory. The classical theory is assumed to be a coarse-grained version of the more fundamental quantum theory, a mere shadow of the richer quantum description. So it is no surprise that generically the classical description cannot uniquely determine the quantum theory. For complex theories there are many ambiguities. In each of these general procedures there are choices to be made. Surely, many of these choices may be physically or mathematically motivated, but a priori, nothing built-in in these procedures guarantees the constructed theory will be the one preferred by nature. Actually, for complex theories like gravity, these general procedures do not even guarantee that the resulting theory will satisfy the correspondence principle and recover the semi-classical limit one started with! This is something that needs to be checked by hand.

We focus now in canonical quantization. As we have reviewed, in the quantum theory as we have described it, a big part of the physical information is contained in the operators and

[9] In a way, this is what was done to construct some pieces of the standard model of particle physics for which we did not have a classical theory.

relations among them. In canonical quantization these relations are postulated or extracted directly from the relations classical observables satisfy (which contain all the information of the classical description). But which are exactly the relevant relations for quantization? A physical justification or motivation comes from the fact that in physical quantum systems certain pairs of observable variables satisfy generalized uncertainty relations which mathematically are the result of their corresponding operators not commuting. Through the correspondence principle, these variables turn out to be canonical pairs in the classical description. This hints at the Poisson algebra structure of the classical observables and the commutator algebra of quantum operators as the relevant algebraic structures to connect both descriptions. Indeed, mathematically, both $(C^\infty(\Gamma), \{\cdot,\cdot\})$ and the quantum algebra of observable operators [as a subset of $(GL(\mathcal{H}), [\cdot,\cdot])$] are Lie algebras. From now on then, for quantization purposes we will extend $C^\infty(\Gamma)$ and consider not only real but also complex-valued functions $f : \Gamma \to \mathbb{C}$. We will keep the same notation and still refer to this extension as the set of classical observable functions[10].

Dirac was first to postulate the *canonical commutation relations* (CCRs) from the Poisson bracket. In fact, he was more ambitious and also proposed a broader correspondence for quantization. In modern and slightly more rigorous terms: The existence of a Hilbert space $\mathcal{H}$ and a *quantization mapping*:

$$\hat{} : C^\infty(\Gamma) \to GL(\mathcal{H})$$
$$f \mapsto \hat{f}$$

satisfying (among other more technical requirements) that:

(i) The map $f \mapsto \hat{f}$ is linear

$$\widehat{\alpha f + \beta g} = \alpha \hat{f} + \beta \hat{g}$$

(ii) Correspondence of Poisson bracket and commutator (Lie algebra morphism)

$$[\hat{f}, \hat{g}] = i\hbar \widehat{\{f, g\}}$$

(iii) Constant function $f(q,p) = 1$ must be mapped to identity operator $\hat{f} = \mathbb{I}$.

(iv) Reality conditions $\hat{f} = \hat{f}^\dagger$.

We now know that for general phase spaces it is impossible for all the algebra of observables $C^\infty(\Gamma)$ to satisfy these conditions consistently. Even for simple cases like $\Gamma = \mathbb{R}^{2n}$, there are "No-go" theorems (e.g. Van-Hove) which point at obstructions to constructing this mapping. So in general we can only choose a proper subalgebra of $C^\infty(\Gamma)$ to be represented unambiguously as quantum operators. This is the first and a key choice in canonical quantization.

3.1. Dirac's procedure

Let us now elaborate on how one can exploit this (partial) correspondence between quantum and classical observables to construct the quantum theory. Canonical quantization may be thought roughly to consist of four steps:

(i) *Choose subalgebra of basic observables*: This is a subset $\mathcal{A} \subset C^\infty(\Gamma)$ of (real or complex) functions on phase space which will correspond to unambiguous quantum operators. This choice is however not arbitrary. At the very least this set $\mathcal{A}$ of basic observables must satisfy:

[10] This is a harmelsss extension since any complex function f may be constructed from two real functions $f = \mathrm{Re}[f] + i\mathrm{Im}[f]$. This extension makes $C^\infty(\Gamma)$ a complex vector space and $(C^\infty(\Gamma), \{\cdot,\cdot\})$ a complex algebra like $(GL(\mathcal{H}), [\cdot,\cdot])$.

- The set $\mathcal{A}$ should be 'controllable'. This usually means $\mathcal{A}$ is closed under the operation of taking the Poisson bracket, i.e. $\mathcal{A}$ should be a Poisson subalgebra of the algebra of observables.
- The set of basic observables should not be too large to preclude the existence of a quantization map, but it should be large enough to *separate points on* Γ. This means, we should distinguish different points on phase space just by the values observables in $\mathcal{A}$ take on these points: $x, x' \in \Gamma$ and $x \neq x'$ implies $\exists f \in \mathcal{A}$ such that $f(x) \neq f(x')$. This guarantees $\mathcal{A}$ contains enough geometric information of the phase space.

Example: In standard quantum mechanics (finite number of degrees of freedom), one fixes canonical coordinates $(q, p) \in \Gamma$ and chooses coordinate functions and the constant unit function. For a single canonical pair:

$$f_0(q,p) = 1, \quad f_1(q,p) = q, \quad f_2(q,p) = p\,.$$

The canonical relations give

$$\{q,p\} = 1, \qquad \text{and} \qquad \{q,q\} = \{p,p\} = \{q,1\} = \{p,1\} = 0\,.$$

Therefore the generated Poisson subalgebra are linear combinations $\mathcal{A} = \{f = a+bq+cp\,|a,b,c \in \mathbb{C}\}$. This is the *Heisenberg algebra.* Clearly the Heisenberg algebra separates points since it contains the coordinate functions.

For the same system, we may however choose different subalgebras of observables:

- One may consider the family of functions

$$F_1(q,p) = q, \quad V_\lambda(q,p) = e^{i\lambda p}, \quad \forall \lambda \in \mathbb{R}$$

  which is closed under the Poisson bracket:

$$\{q, V_\lambda\} = i\lambda V_\lambda, \quad \{q,q\} = \{V_{\lambda_1}, V_{\lambda_2}\} = 0$$

  The generated Poisson subalgebra $\mathcal{A}$ consists of linear combinations of these functions and it also separates points.
- A similar but different choice would be the Poisson algebra generated by the family: $F_1(q,p) = p$ and $U_\mu(q,p) = e^{i\mu q}$, for $\mu \in \mathbb{R}$

These last two choices of basic observable subalgebras $\mathcal{A}$ are analogs of the so called *Holonomy-flux algebra* chosen in LQG. These choices may seem less natural here, but they are well motivated in the gravitational case.

(ii) *Construct the quantum algebra* $\mathfrak{A}$ *of basic observables*: This is the detailed version of 'imposing the canonical commutation relations'. More accurately, what one is doing is:

- Construct abstract (free-associative) algebra of basic operators by promoting basic observable functions in $\mathcal{A}$ to abstract operators: $f \mapsto \widehat{f}$.
- Impose commutation relations

$$[\widehat{f}, \widehat{g}] = i\hbar\widehat{\{f,g\}} \qquad\qquad \forall f, g \in \mathcal{A}\,,$$

- along with an abstract *involution* $* : \mathfrak{A} \to \mathfrak{A}$. Recall this is a map satisfying:
  (a) $(\alpha A + \beta B)^* = \bar{\alpha} A^* + \bar{\beta} B^*$
  (b) $(AB)^* = B^* A^*$
  (c) $(A^*)^* = A$

This involution represents reality conditions. So for real functions $f = \bar{f}$, impose $\widehat{f} = \widehat{f}^*$ (eventually this will correspond to adjoint operation in a concrete Hilbert space)

$\mathfrak{A}$ is called the *quantum $*$-algebra of basic observables*. This is an abstract mathematical object containing the relevant algebraic information of the system. At this point there is no specific Hilbert space or concrete operators yet, only the abstract commutation relations.

Example: In standard quantum mechanics abstract operators $\widehat{x}$ and $\widehat{p}$ generate free algebra of polynomials in $\widehat{x}$ and $\widehat{p}$, and we have the *Canonical Commutation Relations* (CCRs):

$$[\widehat{x}, \widehat{x}] = i\hbar\widehat{\{x, x\}} = \hat{0}\,, \qquad [\widehat{p}, \widehat{p}] = i\hbar\widehat{\{p, p\}} = \hat{0}\,,$$
$$[\widehat{x}, \widehat{p}] = i\hbar\widehat{\{x, p\}} = i\hbar\hat{1}\,. \tag{8}$$

(iii) *Find representations of the quantum $*$-algebra $\mathfrak{A}$*: This is the crucial step of implementing the CCRs, i.e. finding a concrete Hilbert space $\mathcal{H}$ and specific operators that obey the CCRs. More rigorously, one needs to find a (kinematical) Hilbert space $\mathcal{H}$ and a map $\pi : \mathfrak{A} \to GL(\mathcal{H})$ which preserves the $*$-algebra structure:

(a) $\pi(\alpha\widehat{f} + \beta\widehat{g}) = \alpha\pi(\,\widehat{f}\,) + \beta\pi(\,\widehat{g}\,)$

(b) $\pi(\,\widehat{f}\,\widehat{g}\,) = \pi(\,\widehat{f}\,) \circ \pi(\,\widehat{g}\,)$

(c) $\pi(\,\widehat{f}^*\,) = \left[\pi(\,\widehat{f}\,)\right]^\dagger$

But how do we find this Hilbert space and operators? do we have to use our instinct and guess the answer? Luckily, this is now a highly nontrivial but purely mathematical problem in the realm of so called *representation theory.* To try to find representations for the quantum $*$-algebra $\mathfrak{A}$, one can resort to powerful representation theorems and constructions like the Gelfand isomorphism, the GNS-construction (Gelfand-Neimark-Segal), Riesz representation theorem, etc.

Example: For the CCRs (8) Schrödinger gave us the answer. In his *position representation* of standard quantum mechanics:

$$\mathcal{H}_{\text{Sch}} = L^2\,(\mathbb{R}, \mathrm{d}q)$$

and the map $\pi : \mathfrak{A} \to GL(\mathcal{H}_{\text{Sch}})$ represents abstract operators $\widehat{q}$ and $\widehat{p}$ as multiplication an derivation operators acting on square integrable 'wave functions' $\psi \in \mathcal{H}_{sch}$:

$$\pi(\hat{q})\psi(q) := q\,\psi(q)\,, \qquad \pi(\hat{p})\psi(q) := -i\frac{\partial\psi(q)}{\partial q}\,.$$

Of course, to avoid cluttering our formulas, we physicists simply write: $\hat{q}\psi(q) = q\,\psi(q)$ and $\hat{p}\psi(q) = -i\frac{\partial\psi(q)}{\partial q}$ (we have also set $\hbar = 1$).

The well known example above may give the impression that we are overly complicating things. However, for more sophisticated theories –like gravity– it pays off to have a refined or more formal understanding of what 'imposing the commutation relations' really means.

Already for the canonical commutation relations of standard non relativistic quantum mechanical systems of our example, our discussion raises the important question about the uniqueness of the Schrödinger representation. Indeed, by a simple Fourier transformation, we may come up with a 'different' representation of the CCRs. In the *Schrödinger momentum representation* $\mathcal{H} = L^2\,(\mathbb{R}, \mathrm{d}p)$. While the Hilbert space of 'momentum wave functions' is mathematically the same as that of 'position wave functions' (square integrable functions), the representation of each abstract $\widehat{q}$ and $\widehat{p}$ is completely different: For $\psi \in \mathcal{H}$:

$$\pi(\hat{q})\psi(p) = i\frac{\partial\psi(p)}{\partial p}, \qquad \pi(\hat{p})\psi(p) = p\psi(p)$$

Inspired by the ladder operators of the harmonic oscillator:

$$\widehat{a} := \frac{1}{\sqrt{2}}(\widehat{x} + i\widehat{p}), \qquad \widehat{a}^{\dagger} := \frac{1}{\sqrt{2}}(\widehat{x} - i\widehat{p}),$$

we can go further and give another *Fock-type representation* of the CCRs with different Hilbert space $\mathcal{H} = l^2$ of square summable infinite sequences. For any sequence $(z_0, z_1, z_2, \dots) \in l^2$ we may formally define shifting operators:

$$\hat{a}(z_0, z_1, , z_2, \dots) := (z_1, \sqrt{2}\, z_2, \sqrt{3}\, z_3, \dots),$$
$$\hat{a}^{\dagger}(z_0, z_1, z_2, \dots) := (0, z_0, \sqrt{2}\, z_1, \sqrt{3}\, z_2, \dots),$$

and then check $\pi(\widehat{q}) = \sqrt{2}(\widehat{a} + \widehat{a}^{\dagger})/2$ and $\pi(\widehat{p}) = \sqrt{2}(\widehat{a} - \widehat{a}^{\dagger})/2i$ satisfy the CCRs. We can give yet another example. In the *Bargmann-Fock representation* (which is directly obtained if one uses geometric quantization), $\mathcal{H}$ is the Hilbert space of all holomorphic functions $f : \mathbb{C} \to \mathbb{C}$ with inner product

$$\langle \phi_1 | \phi_2 \rangle := \int_{\mathbb{C}} \bar{\phi}_1(z) \phi_2(z) e^{-|z|^2} \mathrm{d}\mu(z),$$

and the ladder operators represented now by multiplication and derivation $\hat{a}\phi(z) := z\phi(z)$ and $\hat{a}^{\dagger}\phi(z) := \frac{\partial}{\partial z}\phi(z)$.

The obvious question now is whether these different representations provide us with the same physical predictions and how do we find this out? The answer is unitary equivalence of representations. More precisely, two representations $(\mathcal{H}, \{\pi(\widehat{f})\}_{\widehat{f} \in \mathfrak{A}})$ and $(\tilde{\mathcal{H}}, \{\tilde{\pi}(\widehat{f})\}_{\widehat{f} \in \mathfrak{A}})$ are unitarily equivalent if we can map one another unitarily, i.e. there is unitary map $\mathbf{U} : \mathcal{H} \to \tilde{\mathcal{H}}$ such that $\tilde{\pi}(\widehat{f}) = \mathbf{U}\, \pi(\widehat{f})\, \mathbf{U}^{-1}$ for all $\widehat{f} \in \mathfrak{A}$. which implies all phyisical quantities we measure: expectation values, transition amplitudes etc. are the same.

The reason we do not generally worry about other representations in our basic quantum mechanics course is because for the finite dimensional mechanical case, 'almost all of them' are unitarily equivalent and predict the same physics. In some very precise sense the Schrödinger representation is unique:

Stone-von Neumann Theorem: Any irreducible, unitary representation of the Weyl algebra (exponentiated version of the CCRs) which is continuous in some sense, is unitarily equivalent to the Schrödinger representation.

For field theories there are no general uniqueness results for the analog of the CCRs. There are actually infinitely many inequivalent 'nice' representations of the CCR. However if we are in Minkowski space and we require the quantum theory to 'inherit its symmetries' then one can single out a representation:

Uniqueness of the Fock Representation in QFT (Segal et. al.): Any irreducible, unitary representation of the Weyl algebra (exponentiated version of the CCRs) admitting a unitary representation of the Poincaré group, is unitarily equivalent to the Fock representation.

In LQG, there is a similar uniqueness result for the representations of the holonomy-flux algebra: the so called *LOST-Fleischhak Theorem.*

(v) *Implement and solve quantum constraints and/or dynamics*: This is the last step of canonical quantization. For simple non relativistic systems this translates to constructing a Hamiltonian operator $\widehat{H}$ and solving the Schrödinger equation. This is not straight forward in general. The Hamiltonian function $H : \Gamma \to \mathbb{R}$ does not generally belong to the basic subalgebra of observables $\mathcal{A}$, so $\widehat{H}$ has to be constructed from operators in the *-algebra $\mathfrak{A}$ of basic operators. This introduces additional quantization ambiguities like the well known *ordering ambiguities* which result because the pointwise product of classical observable

IOP Conf. Series: Journal of Physics: Conf. Series **1010** (2018) 012001 doi:10.1088/1742-6596/1010/1/012001

functions does not have a definite quantum analog (for $f, g \in \mathcal{A}$, the product $\{f, g\}$ has unambiguous quantum representation as a comutator but not fg).
As we have reviewed, for gauge theories like general relativity there is an additional complication. The Hamiltonian theory has constraints C^I associated to gauge symmetries which need to be implemented in the quantum theory to eliminate redundacies and identify the true physical states. To do so, one needs to:

(a) Construct quantum constraint operators $\widehat{C}^I$ from basic operators on kinematical Hilbert space $\mathcal{H}_{\text{kin}}$. Again, this step is not trivial and subject to quantization ambiguities because generically the constraint functions $C^I : \Gamma \to \mathbb{R}$ do not belong to $\mathcal{A}$.

(b) Impose constraints to find gauge invariant states:

$$\widehat{C}^I|\psi\rangle = 0\,,$$

or alternatively implement action of finite gauge transformations $\mathbf{U}_g = e^{iC^I}$ and solve:

$$\mathbf{U}_g|\psi\rangle = |\psi\rangle\,.$$

For the first option there is an important subtlety. If the operators $\widehat{C}^I$ are to consistently represent infinitesimal generators of quantum gauge symmetries, the quantization must be *anomaly-free*, i.e. an analog of the first class contraint algebra (7) must hold for the quantum operators:

$$[\widehat{C}^I, \widehat{C}^J] = \widehat{f_K^{IJ}}\,\widehat{C}^K\,.$$

(c) Find physical Hilbert space $\mathcal{H}_{\text{phys}}$ to start doing physics! This extra step is usually necessary because generically, the gauge invariant states $|\psi\rangle$ solving equations above, may not belong to the Hilbert space $\mathcal{H}_{\text{kin}}$. The kinematical Hilbert space $\mathcal{H}_{\text{kin}}$ (where the constraints have not been imposed) then needs to be replaced or extended somehow to contain the gauge invariant states[11].

4. Gravity

We finally turn to the Hamiltonian formulation of gravity and its (loop) quantization. Since our purpose is only to give an overview of the application to gravity of ideas previously discussed, we will be slightly less rigorous and we will only sketch the general ideas.

Recall that in Einstein's theory of general relativity, *spacetime* is modeled by a four-dimensional manifold $\mathcal{M}$ whose points represent each an *event* (some 'here and now') in the universe. According to Einstein, and as a consequence of the equivalence principle, gravity is a manifestation of the curvature of spacetime. Mathematically, this curvature –which we can physically measure– is given by the so called *Riemann curvature tensor* $R^\rho_{\mu\nu\sigma}$. In turn, the curvature tensor may be reconstructed from a simpler object: the *metric* $g_{\mu\nu}$. The components of $R^\rho_{\mu\nu\sigma}$ are quadratic functions of the metric: $R(g, g^2, \partial g, \partial^2 g)$, which hence acts as some kind of potential for curvature [similarly to potential functions $V(q)$ for conservative forces in mechanics: $F = \partial V/\partial q$, or the magnetic vector potential $\vec{A}$ for the magnetic field: $\vec{B} = \nabla \times \vec{A}$].

[11] This happens already for the simplest mechanical systems like the free particle in one dimension. The plane wave solutions $\psi(q) = e^{ipq}$ are not square integrable, i.e. they do not belong to $L^2(\mathbb{R}, dq)$. To properly accommodate plane wave solutions in the theory one needs to construct a *Gelfand triple* or 'rigged space': $\mathcal{S} \subset L^2(\mathbb{R}, dq) \subset \mathcal{S}^*$, where plane wave solutions are interpreted as 'distributions' or elements in the dual $\mathcal{S}^*$ of a subspace $\mathcal{S}$ of $L^2(\mathbb{R}, dq)$ and which act by creating 'wave packets'.

IOP Conf. Series: Journal of Physics: Conf. Series **1010** (2018) 012001 doi:10.1088/1742-6596/1010/1/012001

The principle of general relativity implies first that any free falling observer is entitled to assume he/she is in an inertial reference frame. Mathematically, this translates to the requirement that at each point p on $\mathcal{M}$ one may find coordinates for which $g_{\mu\nu}$ looks like the Minkowski metric of special relativity: $\eta_{\mu\nu} = \text{diag}(-1, 1, 1, 1)$. In other words, spacetime $(\mathcal{M}, g_{\mu\nu})$ is a four-dimensional pseudo-Riemannian manifold[12].

Secondly, the principle of general relativity also implies that physical laws should be independent of the coordinates an observer uses to describe them. In other words, the laws of physics should be invariant under arbitrary coordinate transformations on $\mathcal{M}$ or equivalently diffeomorphisms $\phi : \mathcal{M} \to \mathcal{M}$ of the spacetime manifold. The equivalence principle combined with this coordinate independence then implies *general covariance* of the theory. This general covariance means equations in general relativity theory must be spacetime tensorial and should only involve the metric –or quantities derived from it like curvature– to describe dynamics of spacetime itself. This entails in particular *background independence* of the theory: when written in tensorial form, no additional preferred or 'background' geometrical structures other than the metric should appear in the equations.

Finally, gravity or the geometry of spacetime is dynamical. 'Free falling observers' or *test particles* in general relativity follow geodesics. This is how curvature manifests itself or influences matter. But in turn, $T_{\mu\nu}$ the energy-momentum tensor of matter in a spacetime shapes curvature according to Einstein's equations:

$$R_{\mu\nu} - \frac{1}{2} R g_{\mu\nu} = 8\pi G T_{\mu\nu} \, . \tag{9}$$

In summary, we may destill three key lessons from general relativity: Gravity is a dynamical gauge theory of geometry of a spacetime $(\mathcal{M}, g_{\mu\nu})$. As such, its formulation must be background independent and invariant under diffeomorphisms, i.e. the group of gauge or local symmetries of the theory must necessarily contain Diff($\mathcal{M}$), the group of diffeomorphisms of $\mathcal{M}$.

4.1. Hamiltonian ADM formulation

General covariance implies there is no preferred notion of time in general relativity. However, as we have reviewed, in a classical canonical or Hamiltonian formulation, a definite notion of a time coordinate t not only plays a crucial role to describe dynamics but is also essential to define generalized momenta or a Legendre transform. Einstein's covariant field equations (9) are not in Hamiltonian form. In order to reformulate general relativity in Hamiltonian form, one first needs to select a 'notion of time' in $\mathcal{M}$. This necessarily breaks *explicit* diffeomorphism invariance and covariance. Nevertheless, at the end of the day, the canonical formulation will still be diffeomorphism invariant and background independent.

One may think the goal of this reformulation is to rewrite general relativity as a theory of fields on a three-dimensional space Σ and evolving in time t. The key idea is to slice spacetime into three-dimensional 'constant time t' hypersurfaces Σ_t (one for each t). The picture is similar to how one slices a loaf of bread. This is a *folation* of spacetime with space-like hypersufaces Σ_t. One may think of the Σ_t's as the simultaneity hypersurfaces of some special observer in $\mathcal{M}$. That this can actually be done necessarily implies certain global properties for $\mathcal{M}$, like having topology $\mathbb{R} \times \Sigma$, but we will not worry about all such technicalities here. The foliation by itself is not sufficient to define a notion of time evolution in $\mathcal{M}$, one further needs a (generally time-like) direction transverse to the foliation in order to map and identify points on different hypersurfaces Σ_t. This is given by an *evolution vector field* t^μ on $\mathcal{M}$ or equivalently by its

[12] Whether one may find coordinates for which $g_{\mu\nu}$ looks like the Minkowski metric not only at p but in a neighborhood of p, depends on whether this neighborhood is *flat*, i.e. on the vanishing of the Riemann curvature tensor $R^\rho_{\mu\nu\sigma} = 0$.

IOP Conf. Series: Journal of Physics: Conf. Series **1010** (2018) 012001 doi:10.1088/1742-6596/1010/1/012001

integral curves. These curves allow one to give precise meaning to the notion of 'fixed point in space' [Fig. 4(a)].

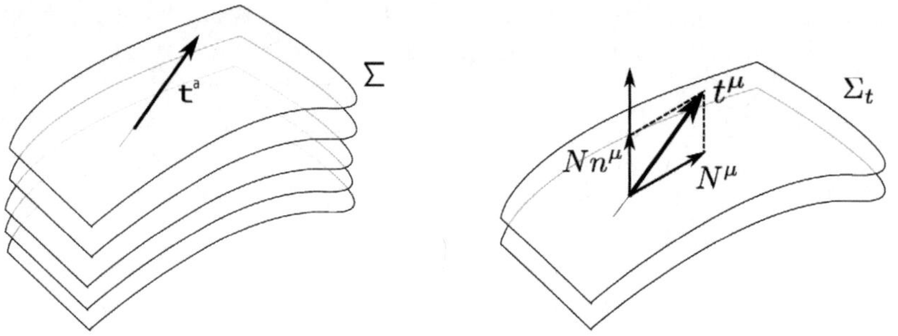

(a) Foliation $\{\Sigma_t\}_{t\in\mathbb{R}}$ and transverse vector field t^μ with integral curves

(b) Decomposition for evolution vector field

Figure 4. Elements for 3+1 decomposition

Using these two geometric background structures, by means of the orthogonal projection of vectors to the hypersurfaces Σ_t, one may decompose every spacetime 4-vector or tensor into a purely spatial part tangent to the foliation and a normal time-like part. This choice of foliation and splitting of tensors is called a 3+1 *decomposition*. For example, for the evolution vector field one has the decomposition:

$$t^\mu = Nn^\mu + N^\mu \,.$$

n^μ is the unit normal to Σ_t, N^μ is the tangential component called the *shift vector*, and N the normal component called the *lapse function* [Fig. 4(b)]. For the metric one has the decomposition:

$$g_{\mu\nu} = q_{\mu\nu} + n_\mu n_\nu \,.$$

The 'tangential' or purely spatial part $q_{\mu\nu}$ defines a three-dimensional Riemannian metric [of signature (+,+,+)] on Σ_t. Every other vector or tensor has a similar decomposition. Now, by following the integral curves of t^μ, one may map one Σ_t to the next and see how the purely spatial or tangential components of tensors change as we move from one Σ_t to the next, i.e. as we 'evolve in time'. It turns out this 'evolution' of three-dimensional fields fully reconstructs the four-dimensional geometry or spacetime dynamics. This identification procedure also defines *adapted coordinates* in $\mathcal{M}$. Given coordinates x^a on a particular Σ_t, by 'dragging' them along t^μ, we may use (t, x^a) as coordinates for $\mathcal{M}$. Using these coordinates the metric is written in block matrix form as

$$g_{\mu\nu} = \begin{bmatrix} -N^2 + N^c N_c & N_b \\ & \\ N_a & q_{ab} \\ & \end{bmatrix}$$

with N^a the 3-dimensional shift vector and q_{ab} the 3×3 matrix representing the spatial metric. Spatial indices are lowered or raised with this 3-metric.

The 3-metric q_{ab} fully describes the *intrinsic* geometry of Σ_t. Additionally, there is another important bilinear form describing the *extrinsic* geometry of Σ_t. This form is the *extrinsic curvature* K_{ab}. The extrinsic curvature describes how Σ_t is embedded in full spacetime $\mathcal{M}$, i.e. how Σ_t bends inside $\mathcal{M}$[13].

[13] Accordingly, extrinsic curvature or 2nd fundamental form is defined as the rate of change or covariant derivative

IOP Conf. Series: Journal of Physics: Conf. Series **1010** (2018) 012001 doi:10.1088/1742-6596/1010/1/012001

One can now derive the Hamiltonian theory. One may start with the covariant Einstein-Hilbert action:

$$\int_{\mathcal{M}} \mathrm{d}^4x \sqrt{-\det g}\, R\,,$$

and perform a 3+1 decomposition to select adapted coordinates (t, x^a) and a notion of time. One can then apply a Legendre transform to define canonical variables and write the action in Hamiltonian form. These are non-trivial steps but we will not go over the details here. The end result is that the induced three-dimensional Riemannian metric $q_{ab}(t, \mathbf{x})$ and the extrinsic curvature $K_{ab}(t, \mathbf{x})$ of Σ_t act as the canonical variables of the theory. More precisely, the momentum density $p^{ab}(t, \mathbf{x})$ conjugate to the metric is $p^{ab} := \sqrt{q}(K^{ab} - q^{ab}K)$ and one has the Poisson bracket canonical relations:

$$\{q_{ab}(\mathbf{x}), p^{cd}(\mathbf{y})\} = 16\pi G \delta^a_{(c}\delta^b_{d)}\delta(\mathbf{x}, \mathbf{y})\,.$$

Einstein's equations in Hamiltonian form are:

$$\begin{aligned} C^a_D &\approx 0, & C_H &\approx 0, \\ \dot{q}^{ab} &= \{q^{ab}, H_{\text{Total}}\}, & \dot{p}_{ab} &= \{p_{ab}, H_{\text{Total}}\} \end{aligned} \tag{10}$$

There are four constraints per point on Σ_t: three are grouped in a vector or spatial *difeomorphism constraint* $C^a_D \approx 0$, $a = 1, 2, 3$, and the remaining one is the scalar or *Hamiltonian constraint* $C_H \approx 0$, their functional form in terms of the canonical variables is of little relevance for our discussion but nevertheless we write it here for completeness:

$$C^a_D := -2D_b p^{ab}\,,$$

$$C_H := \frac{1}{\sqrt{\det q}}\left(q_{ac}q_{bd} - \frac{1}{2}q_{ab}q_{cd}\right)p^{ab}p^{cd} - \sqrt{\det q}\,\mathcal{R}\,.$$

Notice only that the Hamiltonian constraint has the form of a typical Hamiltonian in mechanics: a kinetic term quadratic in momenta p^{ab} plus a (quadratic) potential function $\mathcal{R}(q, q^2, Dq, D^2q)$ of the configuration variables q_{ab}. This potential is not simple though, it is the Ricci scalar of the three dimensional intrinsic curvature.

The total Hamiltonian prescribing evolution in the bulk spacetime turns out to be a sum of the constraints:

$$H_{\text{Total}} = H_{\text{Total}}(q, p, N, N^a) = H[N] + D[N^a]\,,$$

with constraints smeared with lapse and shift:

$$H[N] := \int_\Sigma N\, C_H \qquad \text{and} \qquad D[N^a] := \int_\Sigma N_a\, C^a_D\,.$$

General relativity is hence a totally constrained system, i.e. its total Hamiltonian vanishes on the constraint surface $H_{\text{Total}} \approx 0$. This is a reflection of the general covariance of the theory,

of the normal n^μ to the hypersurfaces Σ_t:

$$K_{\mu\nu} := P^\rho{}_\mu P^\sigma{}_\nu \nabla_\rho n_\sigma\,,$$

where $P^\mu{}_\nu$ denotes the *orthogonal projector* to the hypersurfaces: $P^\mu{}_\nu := \delta^\mu{}_\nu + n^\mu n_\nu$. Extrinsic curvature is purely spatial, so one may use adapted coordinates or abstract spatial indices to denote it: K_{ab}.

IOP Conf. Series: Journal of Physics: Conf. Series **1010** (2018) 012001 doi:10.1088/1742-6596/1010/1/012001

evolution in the bulk in Hamiltonian general relativity is pure gauge. Indeed, the smeared constraints are first classs:

$$\begin{aligned}\{H[M],H[N]\} &= D[q^{ab}(M\partial_b N - N\partial_b M)]\,,\\ \{H[N],D[N^a]\} &= -\,H[N^a\partial_a N]\,,\\ \{D[N^a],D[M^b]\} &= D[[N^a,M^b]]\,,\end{aligned} \tag{11}$$

i.e. their Poisson brackets are linear combinations of the constraints. As we have said before, this implies they are canonical generators of the gauge symmetry transformations of the theory.

As we have emphasized, the gauge symmetries of general relativity encompass spacetime diffeomorphisms. The infinitesimal generators of diffeomorphisms in spacetime are spacetime 4-vector fields V^μ, so that for example, an infinitesimal transformation of the spacetime metric $g_{\mu\nu} \to g_{\mu\nu} + \delta g_{\mu\nu}$ under a diffeomorphism generated by V^μ is given by its Lie derivative along V^μ:

$$\delta g_{\mu\nu} = \epsilon \mathcal{L}_{\vec{V}} g_{\mu\nu}\,, \qquad \text{with} \qquad \epsilon << 1\,.$$

In the 3+1 decomposition for the Hamiltonian theory, one may split this generator into normal and spatial parts:

$$V^\mu = Mn^\mu + M^\mu$$

- The spatial part M^a smeared with the diffeomorphism constraint gives

$$\delta q_{ab} = \epsilon\{q_{ab}, D[M^a]\} = \epsilon\mathcal{L}_{\vec{M}} q_{ab}, \qquad \delta p^{ab} = \epsilon\{p^{ab}, D[M^a]\} = \epsilon\mathcal{L}_{\vec{M}} p^{ab}\,.$$

  This says the smeared diffeomorphism constraint $D[M^a]$ is precisely the canonical generator function on phase space Γ of spatial diffeomorphisms in Σ. Therein its name!

- The normal component M smeared with the Hamiltonian constraint gives

$$\delta q_{ab} = \epsilon\{q_{ab}, H[M]\} \approx \epsilon\mathcal{L}_{M\vec{n}} q_{ab}, \qquad \delta p^{ab} = \epsilon\{p^{ab}, H[M]\} \approx \epsilon\mathcal{L}_{M\vec{n}} p^{ab}$$

  which says the smeared Hamiltonian constraint $H[M]$ is the canonical generator function of diffeomorphisms 'normal' to Σ. These normal diffeomorphisms translate or map Σ_t to $\Sigma_{t+\delta t}$ and hence correspond to 'time evolution' in the Hamiltonian formulation. Therein the name for $H[M]$!

So despite of requiring additional background structures $\{\Sigma_t\}_{t\in\mathbb{R}}$ and t^μ which seems to break explicit diffeomorphism invariance and covariance, the Hamiltonian theory still has Diff($\mathcal{M}$) as part of its gauge symmetry group. How does the theory manage to be independent of the additional background structures as well? The lapse function $N(t,\mathbf{x})$ and shift vector $N^a(t,\mathbf{x})$ parameterize different choices of foliations of spacetime and evolution vector fields. These functions are not dynamical variables and are left completely unrestricted by the Hamiltonian equations (10). So the theory is independent of these background structures because we are free to choose them arbitrarily.

The information that we have a covariant theory is actually completely contained or codified in the Poisson bracket relations (11). These relations are a representation of the *Hypersurface deformation algebra.* This set is a fundamental object, encoding not only the gauge symmetries of Einstein's theory but the structure of spacetime itself. Poisson bracket relations (11) express the fact that dynamics takes place on space-like hypersurfaces embedded in a pseudo-Riemannian manifold (Hojman,Teitelboim,Kuchar). With explicit diffeomorphism invariance and covariance broken, it is then key that these relations be appropriately represented in the corresponding quantum theory.

4.1.1. Wheeler - DeWitt Quantization Given the Hamiltonian ADM formulation of general relativity, one may try to apply Dirac quantization:

(i) The obvious choice for a subalgebra of observables are the very same canonical variables (q_{ab}, p^{ab}) and the Heisenberg-type algebra they generate.
(ii) So one may promote these functions on phase space to abstract operators:

$$q_{ab},\, p^{ab} \quad \longrightarrow \quad \widehat{q_{ab}},\, \widehat{p^{ab}}$$

to construct the abstract quantum algebra.
(iii) One could then guess or try to find a representation where these operators act by multiplication and derivation on a Hilbert space of "square integrable wave functions of the metric: $|\Psi\rangle = \Psi(q_{ab})$".
(iv) Assuming one has constructed such Hilbert space and representation, one may proceed and construct Hamiltonian and diffeomorphism constraint operators $\widehat{H}[N]$ and $\widehat{D}[N^a]$ and solve

$$\widehat{H}[N]|\Psi\rangle = 0 \qquad \text{and} \qquad \widehat{D}[N^a]|\Psi\rangle = 0$$

for the gauge diffeomorphism invariant states.

This is exactly what Wheeler and DeWitt first tried but, except for simplified or reduced models (like cosmological ones), failed to accomplish for full general relativity. At present, there is still no satisfactory construction following this path. One gets stuck with the procedure already at step (iii). Due mainly to the unwieldy nature of the so called *superspace* of spatial metrics, it has proved extremely difficult to rigorously construct such Hilbert spaces of functions of the metric or any other representation for that matter, let alone constraint operators thereon.

A fortunate turn of events came with the realization that if the superspace of metrics is so difficult to handle, then perhaps one is better off doing without it and choosing instead a different set of variables to describe the gravitational field. This is the root of loop quantization.

4.2. Vielbein-connection variables and loop quantization

The gravitational field may indeed be described by a different set of variables: a tetrad e_I^μ and a Lorentz connection $\omega^I_{\mu J}$. Let us take a moment to gain some physical (and mathematical) insight for these alternate variables.

The spacetime metric $g_{\mu\nu}$ acts as a 'potential' for curvature and contains all information about gravity, but there is a different and more fundamental potential coding the same information about the gravitational field. Gravity dictates which observers are 'free falling' and such observers define or carry an *inertial reference frame.* Thus the gravitational field can be viewed as the field that singles out which reference frames are inertial at each point, and further determines how these reference frames at different points are related. This suggests we can use these inertial reference frames as the basic variables to describe gravity.

Mathematically, these preferred reference frames are described by the *cotetrad* one-form field e^I_μ. The cotetrad projects spacetime tangent vector components V^μ in arbitrary frames to its components V^I in an inertial frame (so $I = 0, 1, 2, 3$). One can think every inertial observer has attached to him/her-self an *internal* Minkowski space with Minkowski metric η_{IJ}. The cotetrad then defines a map (actually an isomorphism) from the tangent space at each point to this internal Minkowski space (Fig. 5).

The cotetrad contains all the information about geometry. The spacetime metric is derived from it as

$$g_{\mu\nu} = \eta_{IJ} e^I_\mu e^J_\nu\,.$$

(a) In the standard picture, geometry is extracted from the metric $g_{\mu\nu}$, e.g. $U \cdot V = g_{\mu\nu}U^\mu V^\nu$

(b) More fundamentally, the local geometry is Minkowski and e^I_μ projects components from arbitrary frames to inertial frames, e.g. $U^I = e^I_\mu U^\mu, \quad V^J = e^J_\nu V^\nu$ and $U \cdot V = \eta_{IJ}U^I V^J = (\eta_{IJ}e^I_\mu e^J_\nu)U^\mu V^\nu$

Figure 5. Fundamentally, gravity defines at each point in spacetime inertial reference frames described by a cotetrad field e^I_μ. Comparing with the standard picture, one can deduce the metric field is derived from the cotetrad $g_{\mu\nu} = \eta_{IJ}e^I_\mu e^J_\nu$.

The components of the cotetrad field define a 4×4 matrix. Its inverse e^μ_I is called the *tetrad* and defines a (non-coordinate) orthonormal basis on tangent spaces:

$$g_{\mu\nu}e^\mu_I e^\nu_J = \eta_{IJ}\,.$$

A free falling observer defines or carries not one but an infinite family of inertial reference frames. An inertial observer at a given point in spacetime is free to rotate or boost his/her own frame as he/she pleases, independently of inertial observers at other points. So the cotetrad description of the gravitational field has an additional local 'internal' Lorentz gauge symmetry. Mathematically, a rotated or boosted cotetrad $\tilde{e}^I_\mu := \Lambda^I{}_J e^J_\mu$, for $\Lambda^I{}_J \in SO(1,3)$, determines the same spacetime metric because for a Lorentz matrix $\Lambda^T\eta\Lambda = \eta$:

$$\eta_{IJ}\tilde{e}^I_\mu\tilde{e}^J_\nu = \Lambda^I{}_K\,\eta_{IJ}\,\Lambda^J{}_L\,e^K_\mu\,e^L_\nu = \eta_{KL}e^K_\mu\,e^L_\nu = g_{\mu\nu}\,.$$

On the other hand, the way reference frames at different spacetime points are related to one another is mathematically described by curvature, or more accurately by a *connection*. In general relativity this interconnection among frames is also part of the information the gravitational field provides, so the connection is fully determined by the metric (and therefore also by the cotetrad). Mathematically however, there are many ways to connect or relate frames or vectors on tangent spaces at different points which need not be determined by the frames themselves.

Let us recall a general connection defines a notion of *parallel transport*: given a curve on spacetime from point x to point y, a connection gives a prescription on how to 'translate' vectors from the tangent space at x to the tangent space at y, so one may compare vectors on different tangent spaces. Equivalently then, a connection defines a way to take 'directional derivatives' of vector fields on a manifold: a *covariant derivative* operator ∇_μ (Fig. 6). In general relativity the connection or covariant derivative prescribing parallel transport is the *Levi-Civita connection* compatible with (and hence determined by) the metric: $\nabla_\rho g_{\mu\nu} = 0$. Operationally, the Levi-Civita connection or covariant derivative, is given by the Christoffel 'symbols' $\Gamma^\nu_{\mu\rho}$ which provide a correction to the flat coordinate derivative:

$$\nabla_\mu V^\nu = \partial_\mu V^\nu + \Gamma^\nu_{\mu\rho}V^\rho\,.$$

IOP Conf. Series: Journal of Physics: Conf. Series **1010** (2018) 012001 doi:10.1088/1742-6596/1010/1/012001

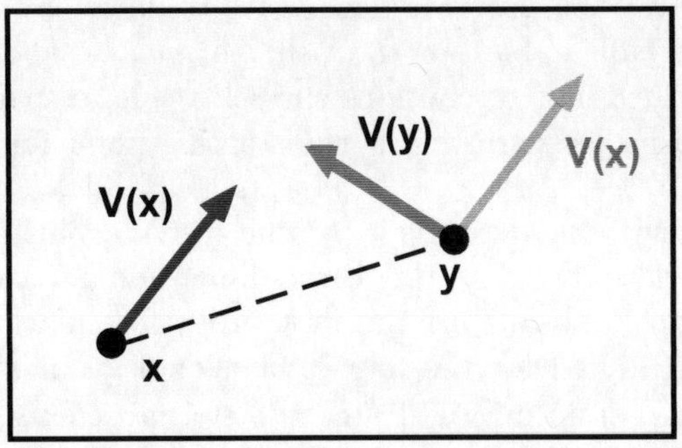

(a) On flat Minkowski or Euclidean space there is a canonical notion of parallel transport to compare vectors at different points. This allows us to define e.g. directional derivatives: $U^a \partial_a V(x) = \lim_{\epsilon \to 0}[V(x + \epsilon U) - V(x)]/\epsilon$.

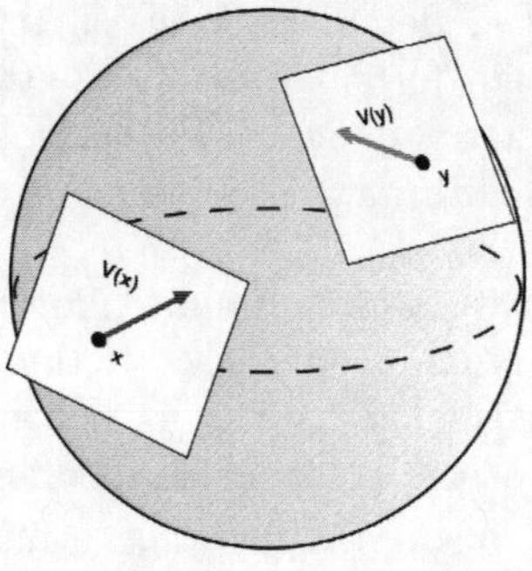

(b) On a general (curved) manifold there is no such preferred notion to compare vectors. These vectors belong to different tangent spaces or more general vector spaces attached at each point. One needs a connection to define derivatives of such vector fields.

Figure 6. General connections on manifolds.

The notion of a connection is actually more general. In the standard metric description of general relativity, one is only concerned with parallel transport and covariant derivatives of vectors V^μ on tangent spaces or their tensor products. But one may define these notions for vectors on more general vector spaces attached to each point of the spacetime manifold[14]. In particular, we have said inertial observers carry an 'internal Minkowski space' defined by their reference frame or cotetrad e^I_μ. To compare vectors V^I on these different internal Minkowski spaces at each spacetime point or –equivalently– to compare or connect different frames e^I_μ, one needs a notion of parallel transport and covariant derivative $\mathcal{D}_\mu$ of internal vector fields V^I, i.e. a connection for the internal spaces. Operationally, this connection or covariant derivative is given by a *connection potential* $\omega^I_{\mu J}$ which 'knows how to act on internal indices' and provides a correction to the flat coordinate derivative:

$$\mathcal{D}_\mu V^I = \partial_\mu V^I + \omega^I_{\mu J} V^J .$$

Physicists usually refer to $\mathcal{D}_\mu$ or $\omega^I_{\mu J}$ simply as 'the connection'. Among the infinitely many possibilities for connections on internal Minkowski space, *Lorentz connections* are defined as those 'compatible with the internal Minkowski metric', i.e. those satisfying $\mathcal{D}_\mu \eta_{IJ} = 0$.

Just like the Levi-Civita connection defines the Riemann curvature $R^\rho{}_{\mu\nu\sigma}$, the connection $\mathcal{D}_\mu$ has an associated curvature $F^I_{\mu\nu J}$.

As we have said, in general relativity the gravitational field e^I_μ itself determines the notion of parallel transport. So the connection $\omega^I_{\mu J}$ must be determined by or be a function of the cotetrad $\omega(e)$. The connection determined by the cotetrad is called the *spin connnection* and it is the one 'compatible with it', in other words it satisfies:

$$\mathcal{D}_\mu e^I_\nu = 0 , \tag{12}$$

and it is necessarily Lorentzian. In a first order formulation of gravity however, the Lorentz connection $\omega^I_{\mu J}$ is not assumed from the outset to be compatible with the cotetrad. This condition becomes a dynamical equation. Both fields $(e^I_\mu, \omega^I_{\mu J})$ then substitute the metric $g_{\mu\nu}$ as the basic dynamical variables to describe gravity.

[14] This defines the notion of general *vector bundles over* $\mathcal{M}$.

After this detour to covariant tetrad-connection variables, we are now ready to describe the Hamiltonian formulation of gravity in terms of so called *Ashtekar-Barbero variables.* These variables actually describe a larger phase space that extends or contains the ADM phase space described with metric variables (q_{ab}, p^{ab}). This alternate description is the starting point for a loop quantization.

In the standard Hamiltonian formulation, gravity is described by the time evolution of Riemannian metric q_{ab} on three-dimensional space Σ. For the three-dimensional (time dependent) geometry (Σ, q_{ab}), there is a description analogous to the first order formulation in terms of the cotetrad and Lorentz connection $(e^I_\mu, \omega^I_{\mu J})$ for the four-dimensional geometry $(\mathcal{M}, g_{\mu\nu})$. In the Hamiltonian formulation, the analog of an inertial frame is a spatial reference frame described by a *cotriad* e^i_a, with $a = 1, 2, 3$, a spatial index and $i = 1, 2, 3$, an 'internal' index. In this case the internal space is good old three dimensional Euclidean space with flat Euclidean metric δ_{ij} =Diag$(1, 1, 1)$, and we have:

$$q_{ab} = \delta_{ij} e^i_a e^j_b \qquad \text{and} \qquad q_{ab} e^a_i e^b_j = \delta_{ij}\,, \tag{13}$$

with e^a_i the inverse orthonormal *triad.* The internal gauge symmetries are $SO(3)$ rotations at each point in Σ. A cotriad or triad can be directly obtained from a cotetrad or tetrad in a 3+1 decomposition of spacetime. Indeed, one can partially reduce the Lorentz gauge freedom of the cotetrad e^I_μ by restricting the frames so that they are 'compatible' with the 3+1 foliation. This basically means that the tetrad basis e^μ_I the spacetime frames define, is such that the time-like basis vector e^μ_0 is normal to Σ_t, and the remaining space-like basis vectors e^μ_i, with $i = 1, 2, 3$, are necessarily tangential and hence can be used as a triad e^a_i. This condition (called the *time gauge*) automatically guarantees (13).

To complete the analogy with four dimensions one needs a connection variable. Just like the cotetrad determines a preferred connection: the four-dimensional spin connection $\omega^I_{\mu J}$ satisfying (12), a cotriad or triad determines a compatible three-dimensional *spin connection* Γ^i_a satisfying

$$D_a e^i_b = 0\,.$$

While the pair (e^i_a, Γ^i_a) fully describes the fixed geometry (Σ_t, q_{ab}) at a given instant of time, in the Hamiltonian formulation of gravity we know the intrinsic geometry of Σ_t defined by the metric q_{ab}, and equivalently by the cotriad e^i_a, is not the complete story. To fully determine the gravitational field one also needs information about the extrinsic geometry of Σ_t, encapsulated in extrinsic curvature K_{ab}. To describe a dynamical or time dependent three-dimensional geometry matching the spacetime of general relativity, one then needs a different connection which incorporates this information about extrinsic geometry and combines it with that given by intrinsic geometry.

In a 3+1 decomposition, one can derive such connection variable as a combination of certain components of the covariant Lorentz connection $\omega^I_{\mu J}$. This connection –which we denote A^i_a– is called the *Ashtekar-Barbero connection.* Unfortunately, the relation between A^i_a and $\omega^I_{\mu J}$ is not as simple geometrically as the relation between the cotriad e^i_a and the cotetrad e^I_μ. This however does not mean A^i_a lacks a geometrical interpretation within the 3+1 Hamiltonian formulation. In the part of the extended phase space which corresponds to the standard ADM formulation, the Ashtekar-Barbero connection is expressed as

$$A^i_a = \Gamma^i_a + \beta K^i_a\,.$$

β is a free constant parameter of the theory called the *Barbero-Immirzi parameter* and $K^i_a = K_{ab} e^b_j \delta^{ij}$.

IOP Conf. Series: Journal of Physics: Conf. Series **1010** (2018) 012001 doi:10.1088/1742-6596/1010/1/012001

It is the pair (e^i_a, A^i_a) that describes the time dependent geometry (Σ_t, q_{ab}) matching general relativity. A Hamiltonian description can then be formulated in terms of canonical variables A^i_a and the *densitized triad*:

$$E^a_i := (\det e) e^a_i = \sqrt{\det q}\, e^a_i \,,$$

with $\det e$ denoting the determinant of the cotriad matrix e^i_a. The pair (A^i_a, E^a_i) satisfies canonical Poisson bracket relations[15]:

$$\{A^i_a(\mathbf{x}), E^b_j(\mathbf{y})\} = 8\pi G\beta\, \delta^i_j\, \delta^b_a\, \delta(\mathbf{x}, \mathbf{y}) .$$

The Hamiltonian equations of general relativity in terms of Ashtekar-Barbero variables are similar to (10), except that there is an additional *Gauss constraint* $C_{\mathcal{G}i} \approx 0$ responsible of implementing the additional $SO(3)$ gauge freedom of the triad-connection formulation:

$$C^a_D \approx 0, \qquad C_H \approx 0, \qquad C_{\mathcal{G}i} \approx 0,$$
$$\dot{A}^i_a = \{A^i_a, H_{\text{Total}}\}, \qquad \dot{E}^a_i = \{E^a_i, H_{\text{Total}}\}$$

with total Hamiltonian prescribing evolution in the bulk spacetime again a sum of the diffeomorphim and Hamiltonian constraints:

$$H_{\text{Total}} = H_{\text{Total}}(A, E, N, N^a) = H[N] + D[N^a] .$$

Just for completeness again, we write the form of the diffeomorphism and Hamiltonian constraints in terms of Ashtekar-Barbero variables:

$$D[N^a] = \frac{1}{8\pi G\beta} \int_\Sigma d^3x\, N^a \left(F^i_{ab} E^b_i - A^i_a C_{\mathcal{G}i}\right) \tag{14}$$

$$H[N] = \frac{1}{16\pi G} \int_\Sigma d^3x\, N \frac{E^a_i E^b_j}{\sqrt{\det E}} \left[\epsilon^{ij}{}_k F^k_{ab} - 2(1+\beta^2) K^i_{[a} K^j_{b]}\right] \tag{15}$$

4.2.1. Loop quantization Finally we sketch the Dirac (loop) quantization program for gravity:

(i) *Choice of subalgebra of basic observables*: Instead of choosing the very same canonical variables (A^i_a, E^b_j) and the Heisenberg-type algebra they generate as basic functions on phase space to quantize

$$A^i_a,\, E^b_j \quad \longrightarrow \quad \widehat{A}^i_a,\, \widehat{E}^b_j \,,$$

the hallmark of a *loop quantization* is to select a different subalgebra of basic observables: the *algebra of holonomies and fluxes*.
As we have said, the connection A^i_a defines a notion of parallel transport. A way to translate vectors V^i given a path $\gamma^a(s)$ in space Σ. Mathematically, the parallel transported vector V^i along the curve $\gamma^a(s)$ from point $\mathbf{x}$ to point $\mathbf{y}$ (Fig. 7) is given by the solution to the parallel transport equation:

$$\frac{\mathrm{d}\gamma^a}{\mathrm{d}s} \mathcal{D}_a V = 0 .$$

[15] More technically, the $SO(3)$ gauge freedom can be extended to (the double covering group) $SU(2)$. The internal indices may be thought as $\mathfrak{su}(2)$ Lie algebra indices. The extended phase space for general relativity is coordinatized by the $\mathfrak{su}(2)$-valued connection $A_a = A^i_a \tau_i$ and the dual $\mathfrak{su}(2)$-valued densitized triad vector $E^a = E^a_i \tau^i$ satisfying the canonical relations and with additional gauge freedom

$$A_a \to g A_a g^{-1} + g\partial_a g^{-1}, \qquad E^a \to g E^a g^{-1},$$

for $g(x^a) \in SU(2)$.

IOP Conf. Series: Journal of Physics: Conf. Series **1010** (2018) 012001 doi:10.1088/1742-6596/1010/1/012001

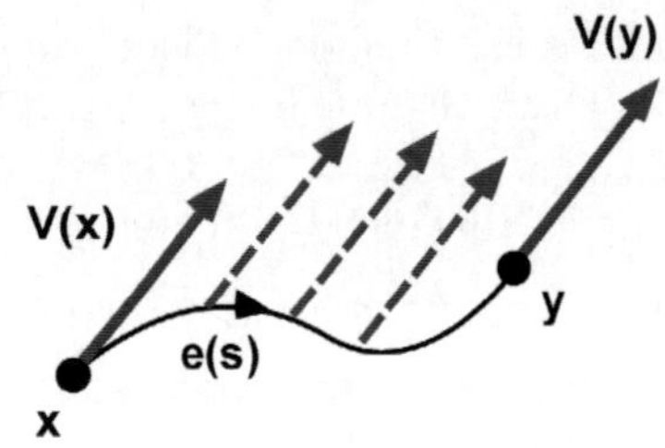

Figure 7. Parallel transport along curve e(s)

This solution can be written as:

$$V(\mathbf{y}) = h_\gamma(A)\, V(\mathbf{x})\,,$$

where $h_\gamma(A)$ is the *holonomy along curve* γ:

$$h_\gamma(A) := \mathcal{P}\exp \int_\gamma A\,.$$

This is essentially the 'exponential of the line integral of the connection A_a^i along γ'. It is a mathematical fact that that if we know the holonomies of a fixed connection for all paths in space we can reconstruct the connection. So holonomies define a special family of functions $\{h_\gamma\}$ on phase space. This family is parameterized by paths $\gamma^a(s)$ on Σ and contains all the geometric information about A_a^i. This family is chosen as a basic subset of observables to quantize.

The other set of basic observables to quantize are the *fluxes* of the densitized triad E_i^a across arbitrary two-dimensonal surfaces $\mathcal{S}$ in space Σ:

$$F_{\mathcal{S},f}(E) := \int_{\mathcal{S}} dS \cdot E_i f^i\,.$$

These surface integrals are the gravitational analogs of electric fluxes in electromagnetism. This family $\{F_{\mathcal{S},f}\}$ is parameterized by 2-surfaces embedded in space Σ and smearing functions f^i on them which are technically necessary to have well defined functions on phase space independent of coordinates and internal bases.

This choice of basic observables may seem odd at first. However it is well justified by its geometric interpretation and by the requirement of background independence[16].

The Poisson brackets of holonomies h_γ and fluxes $F_{\mathcal{S},f}$ have a very neat geometrical description in terms of the paths γ and surfaces $\mathcal{S}$ parameterizing them, and they form a closed subalgebra which further separates points on phase space.

(ii) *Construction of quantum holonomy-flux algebra*: The next step is to promote holonomies and fluxes to abstract operators:

$$h_\gamma(A) \quad F_{\mathcal{S},f}(E) \quad \longrightarrow \quad \widehat{h_\gamma(A)} \quad \widehat{F_{\mathcal{S},f}(E)}\,,$$

and construct an abstract quantum algebra from the Poisson bracket relations:

$$[\widehat{h}_\gamma, \widehat{h}_{\gamma'}] = i\widehat{\{h_\gamma, h_{\gamma'}\}}, \qquad [\widehat{F}_{\mathcal{S},f}, \widehat{F}_{\mathcal{S}',f'}] = i\widehat{\{F_{\mathcal{S},f}, F_{\mathcal{S}',f'}\}}$$
$$[\widehat{h}_\gamma, \widehat{F}_{\mathcal{S},f}] = i\widehat{\{h_\gamma, F_{\mathcal{S},f}\}} \tag{16}$$

[16] Technically, there is no need of some background metric on Σ to properly define the integrals above.

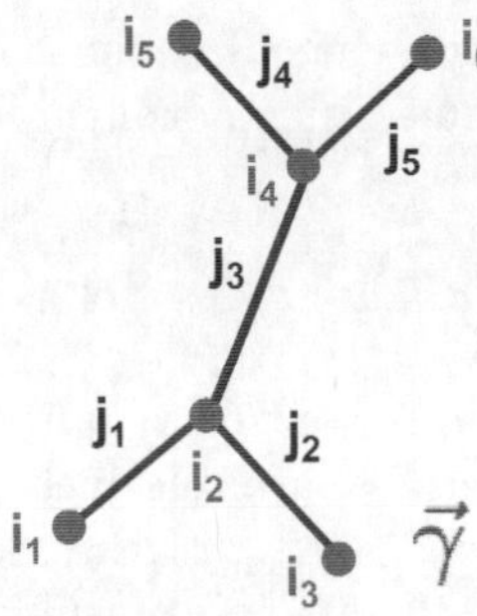

$$\Psi_{\vec{\gamma},\vec{j},\vec{i}}(A) = \prod_{\text{vertices}} i_v \prod_{\text{edges}} \Pi_{j_e}(h_e)$$

Figure 8. Spin networks $\{\Psi_{\vec{\gamma},\vec{j},\vec{i}}\}$ constitute a basis in the kinematical Hilbert space of the loop representation. This basis is labeled by graphs $\vec{\gamma}$ whose edges e (paths) and vertices v are 'colored' with half-integers $j_e \in \{1/2, 1, 3/2, ...\}$ and intertwiners i_v. They are interpreted as basic excitations of geometry.

This is again a nontrivial step, but a rigorous construction was provided by Ashtekar, Corichi and Zapata.

(iii) *Loop representation*: It turns out one can rigorously construct a representation of the (quantum) holonomy-flux algebra (16). The representation may be obtained using the GNS construction and also by using some other equally robust techniques.
The kinematical Hilbert space $\mathcal{H}_{\text{kin}}$ is the space of square integrable complex wave functions $\Psi(A)$ of (generalized) connections:

$$\mathcal{H}_{\text{kin}} = L^2(\bar{A}, \mathrm{d}\mu_{AL}) .$$

In this Hilbert space, holonomies $\widehat{h}_\gamma$ act by multiplication (nonlinear "creation operators"), and fluxes $\widehat{F}_{S,f}$ act by derivation. The subtle part in the construction of the Hilbert space is the definition of a proper (spatial) diffeomorphism invariant *measure* $\mathrm{d}\mu_{AL}$ (the analog of $\mathrm{d}x$) to define integrals for functions on the space of connections $\bar{A}$ which are invariant under changes of coordinates on Σ. One of the most important results of the theory is that this *loop representation* is essentially unique [13]:
LOST-Fleischhak Theorem (Lewandowski, Okolow, Sahlmann, Thiemann, Fleischhak): "Any cyclic representation of the holonomy-flux algebra invariant under spatial diffeomorphisms is unitarily equivalent to the loop representation".
The Hilbert space of the loop representation is non-separable. Still it is quite manegable. In practice, one may use *cylindrical wave functions*. These are special wave functions that depend on the connection A_a^i only through its holonomies along a finite set of 'edges' (paths) composing a 'graph' $\vec{\gamma} = \{\gamma_1, \gamma_2, \ldots, \gamma_N\}$:

$$\Psi_{\vec{\gamma}}(A) = \psi\left(h_{\gamma_1}(A), h_{\gamma_2}(A), \cdots, h_{\gamma_N}(A)\right)$$

In fact using a result from harmonic analysis on compact groups, called the Peter-Weyl theorem, allows to construct an orthonormal basis of *spin networks* (Baez): These are cylindrical functions labeled by graphs $\vec{\gamma}$ embedded in space Σ, with a 'coloring' of edges by half-integer numbers [irreducible representations of $SO(3)$] and another 'coloring' on the vertices where the edges of the graph $\vec{\gamma}$ intersect (this coloring is an assignment of certain matrices called intertwiners). It turns out furthermore, that these basis vectors may be interpreted as polymer-like excitations of geometry (Fig. 8).

(iv) *Dynamics* The last step is to implement dynamics, which for the totally constrained Hamiltonian gravity theory means finding the spacetime diffeomorphism invariant physical states satisfying

$$\widehat{\mathcal{G}}[\Lambda^i]|\Psi\rangle = 0\,, \qquad \widehat{D}[N^a]|\Psi\rangle = 0 \qquad \text{and} \qquad \widehat{H}[N]|\Psi\rangle = 0$$

where $\widehat{\mathcal{G}}[\Lambda^i]$ represents the infinitesimal quantum generator of $SO(3)$ gauge transformations. This operator and its kernel [$SO(3)$ gauge invariant states] are relatively easy to construct in the theory. It turns out that in the full Hilbert space $\mathcal{H}_{\text{kin}}$ of the loop representation one cannot define an infinitesimal generator of spatial diffeomorphisms $\widehat{D}[N^a]$. However on $\mathcal{H}_{\text{kin}}$ there is a unitary representation of the spatial diffeomorphism group Diff(Σ): a diffeomorphism $\phi : \Sigma \to \Sigma$ acts naturally on graphs $\vec{\gamma}$ embedded in Σ by deforming or transforming them as it acts on Σ, so one defines the unitary operator $\mathbf{U}_\phi$ by its action on a cylindrical function

$$\mathbf{U}_\phi \Psi_{\vec{\gamma}}(A) = \Psi_{\phi^{-1}\vec{\gamma}}(A)\,.$$

With this action one can then find, using a technique called *group averaging*, the spatial diffeomorphism invariant states satisfying

$$\mathbf{U}_\phi|\Psi\rangle = |\Psi\rangle \qquad \forall \phi \in \text{Diff}(\Sigma)$$

Finally, given the complexity of the Hamiltonian constraint (15), the implementation of the corresponding operator and the solution of the quantum constraint equation still pose many issues. There are some heroic efforts and impressively, actual candidate operators have been constructed. However, apart from the usual quantization ambiguities for these candidate operators (which are many in this case), the most pressing problem is to guarantee that any of these implementations actually result in a quantum theory free of anomalies which further gives the correct semiclassical limit.

Much work remains to be done in the canonical front of LQG!

Acknowledgments
This work was supported by CONACyT Grant No. 0177840.

References

[1] Perez A 2004 Introduction to loop quantum gravity and spin foams *Preprint* gr-qc/040906
[2] Ashtekar A and Lewandowski J 2004 Background independent quantum gravity: A Status report *Class. Quant. Grav.* **21**, R53
[3] Thiemann T 2007 *Introduction to Modern Canonical Quantum General Relativity Cambridge Monographs on Mathematical Physics* (Cambridge University Press)
[4] Bojowald M 2008 Living Rev. Relativity 11 4. URL: http://www.livingreviews.org/lrr-2008-4
[5] Bojowald M and Skirzewski A 2006 Effective equations of motion for quantum systems *Rev. Math. Phys.* **18** 713-745
[6] Bojowald M, Sandhoefer B, Skirzewski A and Tsobanjan A 2009 Effective constraints for quantum systems *Rev. Math. Phys.* **21** 111-154
[7] Bojowald M and Brahama S 2016 Effective constraint algebras with structure functions *J. Phys. A: Math. Theor.* **49** 125301
[8] Barrau A, Bojowald M, Calcagni G, Grain J and Kagan M 2015 Anomaly-free cosmological perturbations in effective canonical quantum gravity *JCAP* **1505**, no. 05, 051
[9] Reyes J D 2013 Deformed constraint algebras from loop quantum gravity *AIP Conf. Proc.* **1548**, 172
[10] Bojowald M 2010 *Canonical gravity and applications* (Cambrige University Press)
[11] Dirac P A M 1931 *Lectures on quantum mechanics* (Dover Books on Physics)
[12] Henneaux M and Teitelboim C 1992 *Quantization of gauge systems* (Princeton University Press)
[13] Lewandowski J, Okolow A, Sahlmann H, Thiemann T 2006 Uniqueness of diffeomorphism invariant states on holonomy-flux algebras *Commun.Math.Phys.* **267** 703–733.

IOP Conf. Series: Journal of Physics: Conf. Series **1010** (2018) 012002 doi:10.1088/1742-6596/1010/1/012002

Gravity with a non-Abelian gauge symmetry in a Hamiltonian way

Olivera Mišković[1] and Tatjana Vukašinac[2]

[1] Instituto de Física, Pontificia Universidad Católica de Valparaíso, Casilla 4059, Valparaíso, Chile

[2] Facultad de Ingeniería Civil, Universidad Michoacana de San Nicolás de Hidalgo, Morelia, Michoacán 58000, Mexico

E-mail: olivera.miskovic@pucv.cl, tatjana@umich.mx

Abstract. In this note we study possible derivation of new gravitational actions which are also invariant under a non-Abelian gauge symmetry. We start from a canonical representation of symmetry generators and define the corresponding canonical action which can also contain additional Hamiltonian constraints not related to local symmetries. The obtained class of theories possess between zero and some finite number of degrees of freedom. We discuss three special cases, starting with a theory with maximal number of degrees of freedom and finishing with a theory with zero degrees of freedom.

1. Introduction

We are interested in non-Abelian gauge theories which are also invariant under general coordinate transformations, so that they might include gravitational interactions. These kinds of models are interesting from the point of view of coupling the matter fields to gravity, and recently there has been a lot of interest in construction of higher-spin actions. One example of these theories is Chern-Simons gravity for de Sitter, anti-de Sitter or Poincaré gauge group in any space-time dimension.

We start by constructing the canonical representation of the constraints and defining the canonical action as a pure constraint action. This is, by construction, first order action invariant under non-Abelian gauge transformations and spatial or space-time diffeomorphisms. This method was used in two dimensions to obtain Wess-Zumino-Witten action [1] and its supersymmetric version [2]. The method has never been applied in three dimensions. In this paper we address this question, showing that Chern-Simons action is not the only gauge theory which can be constructed.

2. General setup

Consider a gravitational theory that is gauge-invariant under transformations of a semi-simple Lie group G, with the algebra $\mathcal{G}$. For simplicity, we will consider a three dimensional spacetime, with topology $\mathbb{R} \times \Sigma$ and the signature $(-,+,+)$, but our results are easily generalized to an arbitrary dimension.

Let T_a, $a = 1, \ldots, n$, be anti-Hermitean generators of $\mathcal{G}$ with the structure constants $f_{ab}^{\ c}$ and the non-degenerate Cartan metric g_{ab},

$$[T_a, T_b] = f_{ab}^{\ c}\, T_c\,, \qquad g_{ab} = i\mathrm{Tr}(T_a T_b)\,. \tag{2.1}$$

The fundamental field in the theory is a connection $A_\mu^a(x)$, whose canonically conjugated momentum is $\pi_\mu^a(x)$. The spacetime manifold is parametrized by the coordinates $x^\mu = (t, x^i)$ $(i = 1, 2)$. The canonical action is defined as a Legendre transformation of a Hamiltonian density $\mathcal{H}(A, \pi, u)$,

$$I_c[A, \pi, u] = \int dt\, d^2x\, [\dot{A}_\mu^a \pi_a^\mu - \mathcal{H}(A, \pi, u)]\,, \tag{2.2}$$

where $u(x)$ are Hamiltonian multipliers. Due to general covariance, $\mathcal{H}$ is a pure constraint, implying that it is a linear combination of all constraints in the theory. Since the theory is invariant under a gauge group G, there must be n first class constraints $G_a(A, \pi) = 0$ generating a symmetry with the local parameter λ^a, such that

$$\{G_a, G_b'\} = f_{ab}^{\ c}\, G_c \delta\,, \tag{2.3}$$

where we use the short-hand notation $G_a \equiv G_a(t, \vec{x})$, $G_b' \equiv G_b(t, \vec{x}')$ and $\delta \equiv \delta^{(2)}(\vec{x} - \vec{x}')$. We will focus on the case when the theory is invariant under gauge transformations of the form $\delta A_\mu^a = -D_\mu \lambda^a$, where $D_\mu \lambda^a = \partial_\mu \lambda^a + f_{bc}^{\ a} A_\mu^b \lambda^c$. On the other hand, the presence of a time derivative in the transformation law of the fields implies that $\dot{\lambda}^a$ should be treated in the canonical formalism as an independent local parameter, therefore involving an additional set of first class constraints, which we denote by $\bar{G}_a(A, \pi) = 0$, such that their time evolution lead to $G_a(A, \pi) = 0$ [3].

Since the theory should also be invariant under space-time diffemorphisms, there are additional first class constraints $\mathcal{H}_\mu(A, \pi)$, generating transformations with the parameter $\xi^\mu(x)$. Gauge transformations generated by $\mathcal{H}_\mu$ coincide with diffeomorphisms only on-shell [4]. Namely, the diffeomorphism group is a *kinematical* symmetry of any diffeomorphism invariant action. On the other hand, the constraints depend on the explicit form of the Lagrangian, so the infinitesimal transformations they generate form a *dynamical* symmetry group. Furthermore, the explicit non-covariance of the Hamiltonian prescription makes the temporal generator $\mathcal{H}_t(A, \pi)$ difficult to obtain. To simplify the problem, we will consider only generators along spatial coordinates, $\mathcal{H}_i(A, \pi)$. Thus, we have 2ε constraints associated to diffeomorphisms, where $\varepsilon = 1$ (if they are independent on gauge transformations) and $\varepsilon = 0$ (if they are dependent). This generically lead to theories with spatial diffeomorphisms only. We shall discuss this issue later.

There also might exist an even number $2m$ $(m \geq 0)$ of second class constraints $\phi_M(A, \pi) = 0$, not related to any local symmetry.

Taking all constraints into account, that is, $2n + 2\varepsilon$ first class ones and $2m$ second class ones, we obtain the theory with $n - m - 2\varepsilon$ degrees of freedom.

To construct the action with these degrees of freedom, we first need a representation of the constraints.

3. Canonical representation of the generators

We seek for a canonical generator

$$G[\lambda, \epsilon] = \int d^2x\, \left(\lambda^a G_a + \epsilon^a \bar{G}_a\right)\,, \tag{3.1}$$

which produces the gauge transformations through the Poisson brackets,

$$\begin{aligned} \delta A_t^a &= \{A_t^a, G[\lambda, \epsilon]\} = \epsilon^a\,, \\ \delta A_i^a &= \{A_i^a, G[\lambda, \epsilon]\} = -D_i \lambda^a\,, \end{aligned} \tag{3.2}$$

IOP Conf. Series: Journal of Physics: Conf. Series **1010** (2018) 012002 doi:10.1088/1742-6596/1010/1/012002

where $\epsilon^a = -D_t\lambda^a$. One particular solution of the above equations leads to the representations

$$G_a = D_i\pi^i_a\,, \qquad \bar{G}_a = \pi^t_a\,. \tag{3.3}$$

Since $\bar{G}_a$ are linear in momenta, they must be *primary* constraints, which follow directly from the definition of momenta, $\pi = \delta I/\delta\dot{A}$. On the other hand, the constraints G_a could be secondary, or a linear combination of primary and secondary constraints, as it happens in Chern-Simons case [5].

To construct the diffeomorphism generators, we first recall the identity

$$D_\mu\left(\xi^\nu A^a_\nu\right) = \pounds_\xi A^a_\mu + \xi^\nu F^a_{\mu\nu}\,, \tag{3.4}$$

where $F^a_{\mu\nu} = \partial_\mu A^a_\nu - \partial_\nu A^a_\mu + f_{bc}^{\ \ a} A^b_\mu A^c_\nu$ is the field-strength. The above relation shows that the gauge transformations and diffeomorphisms can be dependent on-shell, if the equations of motion are $F^a_{\mu\nu} = 0$ (which is true in a theory with zero degrees of freedom, i.e., when A is a pure gauge). In general, gauge transformations and diffeomorphisms are independent.

From the previous identity we can deduce the form of the spatial diffeomorphisms generators,

$$\mathcal{H}_i = -A^b_i G_b + F^b_{ik}\pi^k_b\,. \tag{3.5}$$

It is straightforward to check that, under spatial diffeomorphisms, A^a_i transforms like a vector, while π^i_a transforms like a vector density of weight one. Direct calculation shows that $\mathcal{H}_i$ close the spatial part of the Dirac algebra [6],

$$\{\mathcal{H}_i, \mathcal{H}'_j\} = \mathcal{H}'_i\partial_j\delta + \mathcal{H}_j\partial_i\delta\,, \tag{3.6}$$

and they commute with the gauge generators,

$$\{G_a, \mathcal{H}'_i\} = G'_a\partial_i\delta\,. \tag{3.7}$$

Since G_b and $\mathcal{H}_i$ are first class constraints, it follows that

$$K_i \equiv F^b_{ik}\pi^k_b = \mathcal{H}_i + A^b_i G_b\,, \tag{3.8}$$

are also first class constraints. If G_a and $\mathcal{H}_i$ are independent, the constraints K_i have the form of primary constraints. Their Poisson brackets are

$$\begin{aligned} \{G_a, K'_i\} &= 0\,, \\ \{K_i, K'_j\} &= K'_i\partial_j\delta + K_j\partial_i\delta - F^a_{ij}G_a\delta\,. \end{aligned} \tag{3.9}$$

Note that the Poisson brackets of K_i do not close without the presence of G_a.

The above discussion implies that, when spatial diffeomorphisms and gauge transformations are independent, we have the following constraint structure:

Primary I class	$\bar{G}_a = \pi^t_a\,,$	$n\,,$
Primary/secondary I class	$G_a = D_i\pi^i_a\,,$	$n\,,$
Primary I class	$K_i = F^a_{ij}\pi^j_a\,,$	$2\varepsilon\,,$
Primary/secondary II class	$\phi_M\,,$	$2m\,.$

The most general Hamiltonian density reads

$$\mathcal{H} = u^a\bar{G}_a + v^a G_a + \zeta^i K_i + w^M\phi_M\,, \tag{3.10}$$

where (u, v, ζ, w) are arbitrary multipliers. It is worthwhile noticing that this is an *extended* Hamiltonian which includes both primary and secondary constraints, obtained from the definition of momenta and evolution of primary constraints, respectively. It is different from the *canonical* Hamiltonian, a Legendre transformation of the Lagrangian, unique only up to constraints. The resulting equations of motion of these two Hamiltonians are not identical, but the difference is not physical.

In order to obtain the canonical action (2.2), the next step is to determine some multipliers in terms of the fields using the consistency conditions of primary constraints. To this end, we have to specify how many secondary constraints we have, and which constrains are primary and which secondary.

4. Construction of the canonical action

In the previous section we found a particular representation of first class constraints and wrote an extended Hamiltonian describing a theory with $n - m - 2\varepsilon$ degrees of freedom. Now we have to choose m and ε where, as we discussed before, we always have $\varepsilon = 1$, except when $F = 0$. Let us focus on special cases.

Case 1. Let us assume that there is a minimal number of constraints, that is, there are no second class constraints and no diffeomorphism constraints. The constraint structure is:

Primary I class	$\bar{G}_a = \pi_a^t\,,$	$n\,,$
Secondary I class	$G_a = D_i\pi_a^i\,,$	$n\,.$

The theory has n degrees of freedom. The Hamiltonian density (3.10) becomes

$$\mathcal{H}_1 = u^a\pi_a^t + v^a D_i\pi_a^i\,. \tag{4.1}$$

A time evolution of the variable $X(A,\pi)$ on the phase space is given by $\dot{X} = \int d^2x'\,\{X, \mathcal{H}'\}$. We also have to ensure that G_a are secondary constraints following from the consistency conditions on $\bar{G}_a$. We find that $\dot{\pi}_a^t = D_i\pi_a^i$ only if $v^a = -A_t^a$. The Hamiltonian is, therefore,

$$\mathcal{H}_1 = u^a\pi_a^t - A_t^a D_i\pi_a^i\,, \tag{4.2}$$

and the corresponding Hamiltonian equations have the form

$$\begin{aligned} \dot{A}_t^a &= u^a\,, & \dot{\pi}_a^t &= D_i\pi_a^i = 0\,,\\ \dot{A}_i^a &= D_iA_t^a\,, & \dot{\pi}_a^i &= -f_{ab}^c A_t^b\pi_c^i\,. \end{aligned} \tag{4.3}$$

Furthermore, from (2.2), the canonical action becomes

$$\begin{aligned} I_1 &= \int d^3x\,\left[\dot{A}_i^a\pi_a^i + A_t^aD_i\pi_a^i + \left(\dot{A}_t^a - u^a\right)\pi_a^t\right]\\ &= \int d^3x\,\left(F_{ti}^a\pi_a^i - \bar{u}^a\pi_a^t\right) + \mathrm{BT}\,, \end{aligned} \tag{4.4}$$

where, in the second line, we redefined an arbitrary multiplier as $\bar{u}^a = u^a - \dot{A}_t^a$, and BT denotes the boundary term. The obtained action is not invariant under general coordinate transformations, what is a consequence of an absence of diffeomorphism constraints. Indeed, $F_{\mu\nu}^a = 0$ is not an equation of motion of this theory so, according to (3.4), gauge transformations are not on-shell equivalent to diffemorphisms.

We should compare this example with a well known case of non-Abelian Yang-Mills theory, since they have the same set of the first class constraints $(\pi_a^t, D_i\pi_a^i)$. The essential difference

IOP Conf. Series: Journal of Physics: Conf. Series **1010** (2018) 012002 doi:10.1088/1742-6596/1010/1/012002

between two theories is that in Yang-Mills theory the Hamiltonian density is not a linear combination of constraints, as in (4.2).

The action (4.4) is gauge invariant but not diffeomorphisms invariant. Vanishing Hamiltonian is, therefore, necessary, but not sufficient condition for a theory to be diffeomorphism invariant. This case, therefore, does not give us a good gravitational action. We need to add the diffeomorphism constraints, which is our next case to consider.

Case 2. Let us add the spatial diffeomorphism constraints, K_i, in the Hamiltonian action. In this settings, there are $2n+2$ first class constraints:

$$\begin{array}{lll} \textit{Primary I class} & \bar{G}_a = \pi^t_a\,, \quad K_i = F^a_{ij}\pi^j_a\,, & n+2\,, \\ \textit{Secondary I class} & G_a = D_i\pi^i_a\,, & n\,. \end{array}$$

The number of physical degrees of freedom is $n-2$. The Hamiltonian density is of the form

$$\mathcal{H}_2 = u^a\pi^t_a + v^a D_i\pi^i_a + \zeta^i F^a_{ik}\pi^k_a\,, \tag{4.5}$$

where again we have to replace $v^a = -A^a_t$ to ensure that the evolution of $\bar{G}_a$ leads to G_a. Equations of motion read

$$\begin{aligned} \dot{A}^a_t &= u^a\,, & \dot{\pi}^t_a &= D_i\pi^i_a = 0\,, \\ \dot{A}^a_i &= D_iA^a_t + \zeta^j F^a_{ji}\,, & \dot{\pi}^i_a &= f_{ac}{}^d A^b_t\pi^i_d + D_j\left(\zeta^j\pi^i_a - \zeta^i\pi^j_a\right). \end{aligned} \tag{4.6}$$

The canonical action becomes

$$I_2 = \int d^3x\left(F^a_{ti}\pi^i_a - \zeta^i F^a_{ik}\pi^k_a - \bar{u}^a\pi^t_a\right) + \mathrm{BT}\,, \tag{4.7}$$

where again we introduced the multiplier $\bar{u}^a$.

This example illustrates that it is possible to have a nontrivial spatial-diffeomorphism invariant gauge theory with dynamical degrees of freedom in three dimensions. However, the time-like diffeomorphisms are still absent. In order to include them, we need some additional ingredients, and one possibility is to consider a theory with second class constraints. We do it in the next example.

Case 3. So far we had examples of theories with maximal number of degrees of freedom (minimal number of symmetries). Let us study another extreme case, where a number of degrees of freedom is zero. This can happen only if the fields are pure gauge, or if $F^a_{\mu\nu} = 0$ is fulfilled dynamically. It implies that the diffeomorphisms should not be an independent symmetry. To have zero degrees of freedom, we need $2n$ second class constraints. The simplest possibility is to introduce primary constraints, linear in π^i_a, what leads to the following constraints:

$$\begin{array}{lll} \textit{Primary I class} & \bar{G}_a = \pi^t_a\,, & n\,, \\ \textit{Secondary I class} & G_a = D_i\pi^i_a\,, & n\,, \\ \textit{Primary II class} & \phi^i_a = \pi^i_a + \mathcal{L}^i_a(A)\,, & 2n\,. \end{array}$$

The Hamiltonian density is

$$\mathcal{H}_3 = u^a\pi^t_a + v^a G_a + w^a_i\,\phi^i_a\,. \tag{4.8}$$

Let us analyze the form of $\mathcal{L}^i_a(A)$. From the requirement that the constraint ϕ^i_a commute with G_a and $\bar{G}_a$ it follows that $\mathcal{L}^i_a(A)$ do not depend of A^a_t. Furthermore, since ϕ^i_a are second class constraints, the matrix $\{\phi^i_a, \phi'^j_b\}$ must be invertible. This matrix is antisymmetric under the simultaneous exchange $(i,a,x) \leftrightarrow (j,b,x')$, so that the simplest choice of an invertible matrix is a constant one with the suitable symmetry, that is, $\{\phi_a^i, \phi_b'^j\} = k\,\epsilon^{ij} g_{ab}\,\delta$, or equivalently

$$-\frac{\partial \mathcal{L}_b'^j}{\partial A_i^a} + \frac{\partial \mathcal{L}_a^i}{\partial A_j'^b} = k\,\epsilon^{ij} g_{ab}\,\delta\,, \tag{4.9}$$

leading to a particular solution $\mathcal{L}_a^i = -\frac{k}{2}\,\epsilon^{ij} A_{aj}$, so that

$$\phi_a^i = \pi_a^i - \frac{k}{2}\,\epsilon^{ij} A_{aj}\,. \tag{4.10}$$

However, it turns out that this choice of second class constraints, is not consistent with the first class character of G_a, because their brackets do not vanish,

$$\{\phi_a^i, G_b'\} = f_{ab}^c \phi_c^i \delta + \frac{k}{2}\,\epsilon^{ij} g_{ab} \partial_j \delta\,. \tag{4.11}$$

In order to recover the vanishing brackets between ϕ and G, we redefine these constraints in the following way,

$$\begin{aligned} \tilde{G}_a &= G_a + h_a\,, \\ \tilde{\phi}_a^i &= \phi_a^i + s_a^i\,, \end{aligned} \tag{4.12}$$

where h_a and s_a^i do not depend on A_t^a in order to commute with π_a^t, and we discard a possibility that they depend on momenta because the constraints are already linear in them. Thus, h_a and s_a^i depend on A_i^a only. Now we require that the constraints of first ($\tilde{G}_a$) and second class ($\tilde{\phi}_a^i$) are separated, i.e., their commutator weakly vanishes. As a result, we find

$$h_a = \kappa\,\epsilon^{ij} \partial_i A_{aj}\,, \qquad s_a^i = \frac{k - 2\kappa}{2}\,\epsilon^{ij} A_{aj}\,, \tag{4.13}$$

where κ is a real constant. Now, the algebra of the constraints becomes

$$\begin{aligned} \{\tilde{G}_a, \tilde{G}_b'\} &= f_{ab}^{\;c}\, \tilde{G}_c\,\delta\,, \qquad & \{\tilde{\phi}_a^i, \tilde{G}_b'\} &= f_{ab}^{\;c} \tilde{\phi}_c^i\,\delta\,, \\ \{\tilde{\phi}_a^i, \tilde{\phi}_b'^j\} &= k\,\epsilon^{ij} g_{ab}\,\delta\,. \end{aligned} \tag{4.14}$$

It turns out that the result does not depend on the choice of κ in the redefinition of the constraints (4.13) because it is equivalent to a redefinition of the indefinite multipliers v^a and w_i^a, so we can choose $k = 2\kappa$ without loss of generality, so that $s_a^i = 0$. To summarize, the constraints are

$$\begin{aligned} \tilde{G}_a &= D_i \pi_a^i + \kappa\,\epsilon^{ij} \partial_i A_{aj}\,, \\ \tilde{\phi}_a^i &= \phi_a^i = \pi_a^i - \kappa\,\epsilon^{ij} A_{aj}\,. \end{aligned} \tag{4.15}$$

The consistency condition of the primary constraints lead to $v^a = -A_t^a$ and $w_i^a = 0$, and the Hamiltonian density takes the form

$$\mathcal{H}_3 = u^a \pi_a^t - A_t^a \left(D_i \pi_a^i + \kappa\,\epsilon^{ij} \partial_i A_{aj}\right)\,. \tag{4.16}$$

Hamiltonian equations become

$$\begin{aligned} \dot{A}_t^a &= u^a\,, \qquad & \dot{\pi}_a^t &= D_i \pi_a^i + \kappa\,\epsilon^{ij} \partial_i A_{aj} = 0\,, \\ \dot{A}_i^a &= D_i A_t^a\,, \qquad & \dot{\pi}_a^i &= -f_{ab}^{\;c} A_t^b \pi_c^i + \kappa\,\epsilon^{ij} \partial_j A_{ta}\,. \end{aligned} \tag{4.17}$$

IOP Conf. Series: Journal of Physics: Conf. Series **1010** (2018) 012002 doi:10.1088/1742-6596/1010/1/012002

We also note that $F^a_{ti} = \dot{A}^a_i - D_i A^a_t = 0$ and $F^a_{ij} = \frac{1}{2\kappa}\,\epsilon_{ij}(\tilde{G}^a - D_k\phi^{ka}) = 0$, which means that on-shell $F^a_{\mu\nu} = 0$, as expected, because the theory has zero degrees of freedom. In consequence, the spatial diffeomorphisms are dependent on the gauge transformations, as we saw in (3.4).

Using the equations for the multipliers v^a, w^a_i and the constraints ϕ^i_a to eliminate the non-physical momenta π^i_a in the canonical action (2.2), we obtain

$$I_3 = \int d^3x \left[\kappa\epsilon^{ij}\left(\dot{A}^a_i A_{aj} + A^a_t F_{ij\,a}\right) - \bar{u}^a\pi^t_a\right] . \tag{4.18}$$

We recognize the first part of this action as the Chern-Simons theory,

$$I_{\rm CS}[A] = \kappa \int d^3x\, \epsilon^{\mu\nu\rho}\left(A^a_\mu\partial_\nu A_{a\rho} + \frac{1}{3}\, f_{abc}\, A^a_\mu A^b_\nu A^c_\rho\right) , \tag{4.19}$$

with $\epsilon^{ij} \equiv \epsilon^{tij}$, so that

$$I_3 = I_{\rm CS}[A] - \int d^3x\, \bar{u}^a\pi^t_a \,. \tag{4.20}$$

The actions I_3 and $I_{\rm CS}$ are physically equivalent, corresponding to the extended and canonical Hamiltonian, respectively, as discussed at the end of the previous section. Indeed, the term $\bar{u}^a\pi^t_a$ does not change a space of solutions because $\pi^t_a = 0$, $\bar{u}^a = 0$ on-shell in both theories, and they both have zero degrees of freedom. We can therefore say that $I_3 = I_{\rm CS}$.

5. Conclusions

We used Hamiltonian formalism to explore a physical content of a pure constraint action invariant under non-Abelian gauge symmetries and spatial diffeomorphisms. Our primary motivation was to have a gravitational theory coupled to non-Abelian matter and (or) higher spin fields.

The construction was based on a canonical representation of symmetry generators. Due to non-covariance of the formalism, time-like diffeomorphisms were not included in the construction, which, in one of the examples analyzed (Case 2), led to a non-Abelian theory invariant only under spatial diffeomorphisms. Although this model could not describe a gravity theory, it is remarkable that it possesses $n - 2$ degrees of freedom and is not equivalent to Yang-Mills theory. This theory should be better explored in future. In second example (Case 1), the diffeomorphisms constraints were absent and the theory was not invariant under general coordinate transformations. In the third example (Case 3), the time-like diffeomorphisms were not independent on gauge transformations, leading to a fully diffeomorphism-invariant theory, or Chern-Simons model.

Let us mention that a standard Dirac Hamiltonian analysis of Chern-Simons gravity starts from the known action and arrives to so-called total Hamiltonian which contains only primary constraints (see Chapter 6.4 in Ref.[7]). In our approach, however, we start from the extended Hamiltonian that includes *all* constraints, both primary and secondary, and reconstruct the Chern-Simons action in a different, but physically equivalent way, finding that it is the *unique* action which can be obtained under given assumptions.

We, therefore, showed that Hamiltonian method can be successfully applied to obtain a diffeomorphism invariant theory possessing non-Abelian gauge symmetry.

One of the open questions to be addressed in future is how to treat time-like diffeomorphisms in this formalism in a general way. Another of the future tasks is to analyze the uniqueness of canonical representations, i.e., discuss non-equivalent representations of the generators. The challenge is also to construct a physically interesting example of a theory that possess degrees of freedom, but is also invariant under space-time diffeomorphisms.

Acknowledgments
The authors would like to thank Max Bañados, Sergio Hojman and Jorge Zanelli for useful comments and discussions. This work was supported by CIC, UMSNH.

References
[1] Sazdović B 2000 Canonical approach to 2D WZWN model, non-abelian bosonization and anomalies *Phys. Rev.* D **62** 045011 (*Preprint* hep-th/9912216)
[2] Mišković O and Sazdović B 2002 Canonical approach to 2D supersymmetric WZNW model coupled to supergravity *Mod. Phys. Lett.* A **17** 1923 (*Preprint* hep-th/0111162)
[3] Castellani L 1982 Symmetries in constrained Hamiltonian systems *Annals Phys.* **143** 357
[4] Thiemann T 2007 *Modern canonical quantum general relativity* (New York: Cambridge University Press)
[5] Bañados M, Garay L J and Henneaux M 1996 The dynamical structure of higher dimensional Chern-Simons theory *Nucl. Phys.* B **476** 611 (*Preprint* hep-th/9605159)
[6] Kuchař K V and Stone C L 1987 A canonical representation of spacetime diffeomorphisms for the parametrised Maxwell field *Class. Quantum Grav.* **4** 319
[7] Blagojević M 2002 *Gravitation and gauge symmetries* (Bristol: IoP Publishing)

IOP Conf. Series: Journal of Physics: Conf. Series **1010** (2018) 012003 doi:10.1088/1742-6596/1010/1/012003

Use of Genetic Algorithms to solve Inverse Problems in Relativistic Hydrodynamics

F. S. Guzmán and J. A. González

Instituto de Física y Matemáticas, Universidad Michoacana de San Nicolás de Hidalgo. Edificio C-3, Cd. Universitaria, 58040 Morelia, Michoacán, México.

E-mail: guzman@ifm.umich.mx, gonzalez@ifm.umich.mx

Abstract. We present the use of Genetic Algorithms (GAs) as a strategy to solve inverse problems associated with models of relativistic hydrodynamics. The signal we consider to emulate an observation is the density of a relativistic gas, measured at a point where a shock is traveling. This shock is generated numerically out of a Riemann problem with mildly relativistic conditions. The inverse problem we propose is the prediction of the initial conditions of density, velocity and pressure of the Riemann problem that gave origin to that signal. For this we use the density, velocity and pressure of the gas at both sides of the discontinuity, as the six genes of an organism, initially with random values within a tolerance. We then prepare an initial population of N of these organisms and evolve them using methods based on GAs. In the end, the organism with the best fitness of each generation is compared to the signal and the process ends when the set of initial conditions of the organisms of a later generation fit the Signal within a tolerance.

1. Introduction

In high energy astrophysics, given the experiments cannot be repeated, the explanation and modeling of an event requires the solution of inverse problems, that is, one is interested in tracking the model and the initial conditions of an event, that fit the data of observations within certain error bars.

Simple type of observational data are time-series. These are collection of time dependent data, for instance the light curve of an astrophysical event in a certain frequency band. The models to explain -for instance- gamma ray bursts are of great variety [1], including various ingredients of matter, radiation, interaction between matter an radiation, models in one, two and three dimensions, etc. Each model attempts to solve an inverse problem, with certain parameters that are always faced with data.

The complexity of solving an inverse problem is strongly correlated to the simplicity of the model. For instance, when a set of experimental data has to be fit with a given function with d parameters, the problem reduces to minimizing the error function defined in a parameter space of dimension d. However, more complicated data modeling depends on the solution of Ordinary Differential Equations or Partial Differential Equations (PDEs). For instance, in [2], a model of long gamma ray bursts is given as a solution of the Radiation Relativistic Hydrodynamics equations for a fine tuned set of initial conditions and a highly relativistic shock. This means that instead of trying to fit data with a given function, the data are assumed to be the result of the solution of a complicated system of non-linear PDEs that depends on some initial conditions.

A case illustrating how complicated an inverse problem can be, is the inverse problem of paramount current interest after the Gravitational Wave detection by the LIGO [3]. This problem is the reconstruction of the parameters of a binary black hole system out of the observed Gravitational Wave signal [4, 5].

Given the complexity of some inverse problems, more sophisticated and efficient methods are required that can be used with the current hardware technology. This is the case of GAs. The GA method is an abstraction of biological evolution for moving from one population of organisms (chromosomes, in our case chains of numbers) to a new generation of organisms by using a kind of natural selection together with the genetics inspired operations of crossover and mutation. Each chromosome is made of genes (in our case the pressure, density or velocity of a state of a relativistic gas at initial time). The method has many applications, and for the purposes of the present paper we apply it for an optimization process, that is, the rules of evolution and crossover will be defined in terms of how optimal the genes of a DNA organism are at minimizing an error function, namely the difference between the observed and computed signal.

The problem we choose to test GAs as inverse problem solvers is the solution of a relativistic hydrodynamics shock-tube that we describe below.

2. Numerical methods

2.1. Numerical solution of Relativistic Euler equations

The Riemann problem is an initial value problem for a gas with discontinuous initial data, whose evolution is ruled by Euler's equations. The set of Relativistic Euler's equations determine the evolution of the density of gas, its velocity field and either its pressure or total energy. The model we use is that of a perfect fluid, described by the stress energy tensor

$$T^{\mu\nu} = \rho_0 h u^\mu u^\nu + p\eta^{\mu\nu}, \tag{1}$$

where ρ_0 is the rest mass density of a fluid element, u^μ its four velocity, p the pressure, $h = 1 + \varepsilon + p/\rho_0$ is the specific enthalpy and $\eta^{\mu\nu}$ are the components of the metric describing Minkowski space-time.

The set of relativistic Euler equations is obtained from the local conservation of the rest mass and the local conservation of the stress energy tensor of the fluid, which are respectively

$$\begin{aligned} (\rho_0 u^\mu)_{,\mu} &= 0, \\ (T^{\mu\nu})_{,\nu} &= 0, \end{aligned}$$

where $u^\mu = W(1, v^x, 0, 0)$ and $W = \frac{1}{\sqrt{1-v^i v_i}}$ is the Lorentz factor and v^x is the Eulerian velocity of the fluid elements. It is possible to arrange these equations as a flux balance set of equations as [6, 7]

$$\partial_t \mathbf{u} + \partial_x \mathbf{F}(\mathbf{u}) = 0, \tag{2}$$

where conservative variables are defined by $\mathbf{u} = (D, S^x, \tau)^T$ and the resulting fluxes are $\mathbf{F} = (Dv, Sv + p, S)$, where we assume that specifically $v = v^x$ and $S = S^x$, since we are only considering one spatial dimension. The conservative variables are defined in terms of the primitive ones (ρ_0, v, p) as follows

$$\begin{aligned} D &= \rho_0 W, \\ S &= \rho_0 h W^2 v, \\ \tau &= = \rho_0 h W - p. \end{aligned} \tag{3}$$

In terms of these variables the flux balance equations are explicitly [6, 7]

$$\partial_t D + \partial_x (Dv) = 0, \tag{4}$$
$$\partial_t S + \partial_x (Sv + p) = 0, \tag{5}$$
$$\partial_t \tau + \partial_x S = 0. \tag{6}$$

We assume a Γ−law Equation of State, $p = \rho_0 \varepsilon(\Gamma - 1)$, where Γ is the adiabatic index $\Gamma = c_p/c_v$. The initial data of the Riemann problem is defined by

$$\mathbf{u} = \begin{cases} \mathbf{u}_L, & x < x_0 \\ \mathbf{u}_R, & x > x_0, \end{cases} \tag{7}$$

where $\mathbf{u}_L$ and $\mathbf{u}_R$ represent the values of the gas properties on a chamber at the left and at the right from an interface between the two states at $x = x_0$ that exists only at initial time.

The domain used to prepare the initial conditions and track the evolution is $x \in [0, 1]$, $t \geq 0$ and the interface is intially located at $x_0 = 0.5$. We prepare the initial conditions on the left and right through the primitive variables (ρ_L, v_L, p_L) and (ρ_R, v_R, p_R), where for simplicity we have omitted the subscript zero that identifies the rest mass density.

2.2. Generation of the signal

What we call an observation is actually a numerical solution of the above system, that will be unknown to the GA. The parameters we choose are those of a standard tests for a mildly relativistic shock ($\rho_L = 10, v_L = 0, p_L = 13.33$) and ($\rho_R = 1, v_R = 0, p_R = 0$), using the value $\Gamma = 5/3$ for the adiabatic index. The shock wave moves toward the right of the domain as seen in the snapshots of Fig. 1. We choose the observed signal $Signal(t)$ to be the value of the density measured with a detector at $x_d = 0.75$ as a function of time. This a is pretty close example of how a scalar quantity is measured during a simulation at a given point, for instance in [2] this is how the loss of radiated energy is estimated during the propagation of a strong shock with radiation.

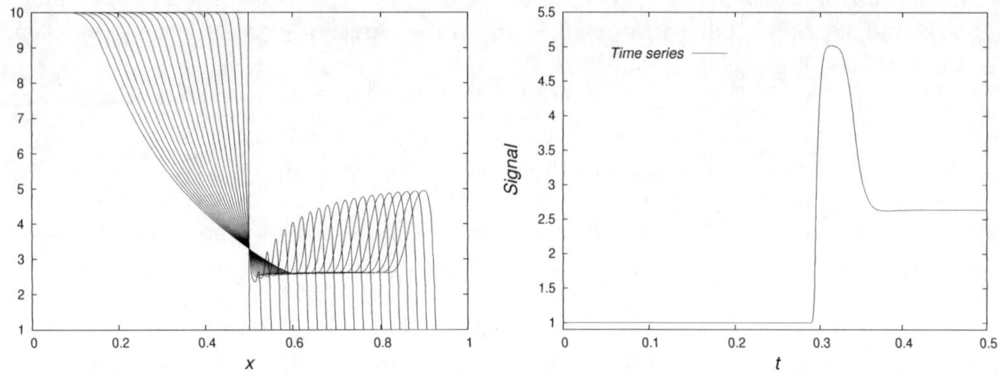

Figure 1. On the left we show the density, first it is discontinuous and once the system starts evolving a shock moves to the right with a mild velocity of 75% the speed of light. On the right we show the value of the density at $x_d = 0.75$ measured in time.

We calculate this numerical solution using a High Resolution Shock Capturing method, with a method of lines that uses the Harten-Lax-van Leer-Einfeldt flux formula, minmod reconstructor and third order Runge-Kutta time integrator (see for instance [8, 9, 10] for some applications).

2.3. Genetic algorithm
The inverse problem we want to solve is the problem of the cause. That is, we consider that the system generating the signal is ruled by the relativistic Euler equations above and that the initial conditions are those of a Riemann problem. However the cause is the set of initial conditions that produces $Signal(t)$. Physically, the observer only knows that the signal was generated by a Riemann problem of the relativistic 1D Euler equations and wonders about the initial conditions that generated the signal.

We define the DNA of an organism as a vector whose entries are the initial conditions. We say the organism i is $(\rho_L, v_L, p_L, \rho_R, v_R, p_R)_i$, where each entry plays the role of a gene.

Then we define a population of organisms $i = 1, ..., N$, that is, we choose N sets of initial conditions that define the DNA of each organism and run a simulation for each of them. We define the fitness of organism i as $F_i = 1/L_1(E_i)$, where $E_i = (\rho(x_d = 0.75) - Signal(t))_i$ is the error of the observed density at $x = 0.75$ with respect to $Signal(t)$ and L_1 indicates the norm one.

For this paper we select the best fit $N/2$ organisms that we keep for the next generation. We complete such new generation with $N/2$ organisms resulting from the crossover of these well fitted $N/2$ organisms. The crossover between organisms is as follows, we take the first three values of the DNA of the organism i and join them to the last three values of the organism j to produce the child $(\rho_L, v_L, p_L, 0, 0, 0)_i + (0, 0, 0, \rho_R, v_R, p_R)_j$. To this new generation we apply a random mutation that may spontaneously give a stronger DNA. Then we run a new set of N simulations and diagnose the fitness, then crossover again and repeat during N_G generations.

2.4. Preparation of the populations
The entries of the organisms can in principle be arbitrary and generated randomly. In practice one knows at least the order of magnitude of the parameters to be found or classified in an inverse problem, in our case here, these are the initial conditions that gave rise to the Time Series. Since we generated the Time Series with initial conditions we know, we can systematize the efficacy of the GA by allowing the values of the initial entries of each organism to have random values within a range around the correct initial condition. We present two cases, one in which we allow the initial entries to have random values within a band of ± 20 and $\pm 30\%$ from the correct initial conditions and a special case where half of the genes are selected randomly within $\pm 50\%$ and the other half within $\pm 10\%$. In all the three cases we set the population to $N = 10$.

3. Results
In Fig. 2 we present the results for the three cases described above. We show the signal detected for the best fitted organism of the first generation, and the best organism of a given latest generation depending on the case, 300, 100 and 300 generations. This illustrates how the initial data of the best initial conditions provides a significantly different signal compared to $Signal(t)$ and how the best fit organism of a later generation fits better the data of $Signal(t)$. We also show how the error of the best fit organism decreases with the number of generations and how it reaches a minimum, after which neither the mutation nor the evolution can improve the fitting. We say that at this stage the DNA of the best fitted organism of the last generation has genes pretty close to the initial conditions that generated the observed $Signal(t)$.

4. Final comments
We have presented a first exploration of the use of GAs in the solution of inverse problems involving PDEs, already similar to those related to realistic long gamma ray burst sources.

The computing time seems to limit the type of problems that can be tackled using this approach. For instance, one of the cases using 300 generations with $N = 10$ organisms each,

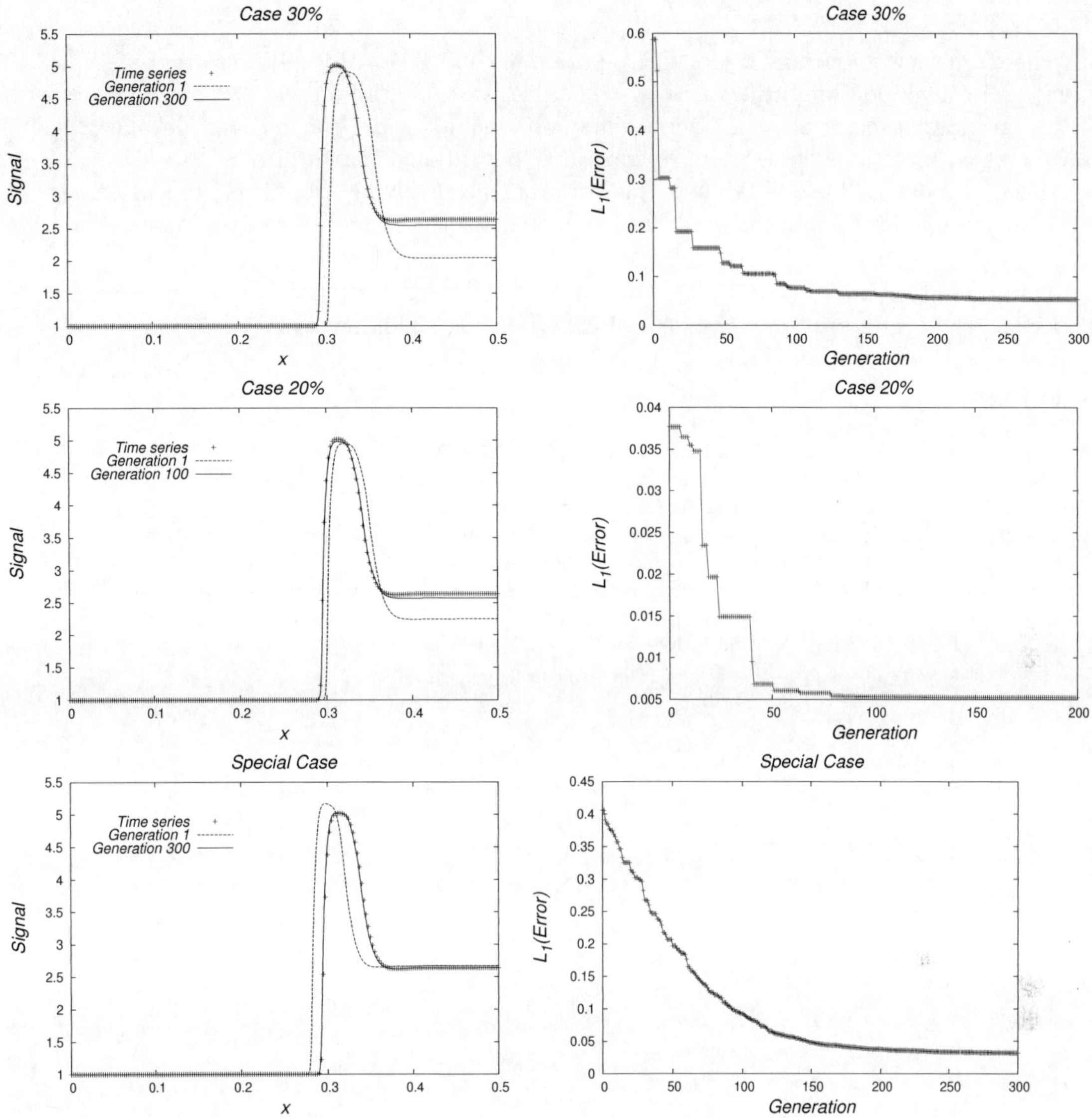

Figure 2. Results for the three different scenarios discussed in this paper. On the left there is the value of the density measured at $x_d = 0.75$ for the organism with the best fitness of the first and the latest generation, and compare it with $Signal(t)$. On the right we show the difference between the best fit organism of each generation using the L_1 norm of the error.

requires the execution of 10 simulations per generation, that is, 3,000 simulations of a shock-tube evolution. The problem uses only one spatial dimension, and with our methods for solving the Relativistic Euler system, each simulation takes about 2 seconds in average, including the GA analysis. This means that the whole process takes about 6,000 seconds, less 2 hours.

In a more realistic scenario, considering more than 6 genes, with a less educated restriction on the initial values of the genes (in our cases within 20%, 30% and a combination of 50% and 10%), it might take a considerably longer number of generations and possibly a bigger population would be needed. Another potential and important limitation is also that increasing the number of dimensions, say from one to two, will increase the number of calculations in orders of magnitude, and therefore the computing time of each solution of the PDE syatem. These are cases in which either more involved computing methods, like the use of parallel code or programming with GPUs are needed, or more sophisticated methods including pattern recognition and machine learning need to be applied as we have done for Gravitational Waves [5].

At the moment, we have shown here that for problems involving one spatial dimension our GA method is usable in the solution of inverse problems associated realistic astrophysical scenarios.

5. Acknowledgments

This research is supported by grants CIC-UMSNH 4.9, 4.23 and CONACyT 258726.

References

[1] P. Kumar and B. Zhang, Phys. Rep. 561, 1-109 (2015).
[2] F. J. Rivera-Paleo and F. S. Guzmán, MNRAS 459, 2777-2786 (2016)
[3] B. P. Abbott et al., Phys. Rev. Lett. 116, 061102 (2016).
[4] J. Veitch, et al. Phys. Rev. D 91, 042003 (2015).
[5] M. Carrillo, M. Gracia-Linares, J. A. González, F. S. Guzmán, Gen. Rel. Grav. 48, 141 (2016)
[6] J. M. Martí and E. Müller, Living Rev. Relativ. (2003) 6: 7.
[7] F. D. Lora-Clavijo, J. P. Cruz- Pérez, F. S. Guzmán, J. A. González. Rev. Mex. Fis. E 59, 28-50 (2013).
[8] M. Gracia-Linares and F. S. Guzmán, ApJ 812, 23 (2015).
[9] M. Gracia-Linares and F. S. Guzmán Phys. Rev. D 94, 064077 (2016).
[10] J. J. González Avilés, F. S. Guzmán, V. Fedun ApJ 836, 24 (2017)

IOP Conf. Series: Journal of Physics: Conf. Series **1010** (2018) 012004 doi:10.1088/1742-6596/1010/1/012004

Equilibrium initial data for luminous matter on top of a BEC dark matter halo

J. A. González and F. S. Guzmán

Instituto de Física y Matemáticas, Universidad Michoacana de San Nicolás de Hidalgo. Edificio C-3, Cd. Universitaria, 58040 Morelia, Michoacán, México.

E-mail: gonzalez@ifm.umich.mx, guzman@ifm.umich.mx

Abstract. Considering dark matter is a Bose Einstein Condensate modeled with the Gross-Pitaevskii equation, and assuming luminous matter is an n-body system, we construct initial data for the luminous matter on top of a BEC dark matter halo. The conditions on the luminous matter are that the peculiar velocities are random, that the density profile is nearly spherical and that the configuration of luminous plus dark matter is a long living system. We analyze the life-time using the full numerical evolution of the configuration.

1. Introduction

The Bose Einstein Condensate (BEC) dark matter model has recently acquired more attention, specially due to the recent advances related to structure formation simulations, under the different name of wave dark matter [1]. In the past, some interesting properties of BEC dark matter were analyzed, specially its properties at galactic core scale helping to solve the cusp-cure problem shown by the more popular CDM model [2, 3, 4, 5]. The main idea of the model is that it assumes dark matter is a condensate of ultralight spinless bosons that can potentially solve the two basic problems of CMD, namely, the cusp-core problem and the over abundance of small structures [1].

In this contribution we focus on the role of the BEC at local scales, specifically in galaxies. It has been very common during the development of the BEC dark matter model, that the interaction of dark and luminous matter has been underestimated. In the first versions of the model, considering it was an exploratory phase of the hypothesis, the main problem was that of the galactic rotation curves (RCs) (see e.g. [6]). Some very strong hypotheses were assumed, for instance, that luminous matter are test particles and travel on circular trajectories. These hypotheses looked reasonable from a pure theoretical point of view, however in practice these are very strong idealizations. On the one hand, there have not been any arguments supporting the first of the hypothesis, it is well known that the process of RCs measurements involves a highly non-trivial data averaging, and that RCs are not measured for exactly symmetric galaxies with the correct orientation as to observe the peculiar red and blue shift of the gas. On the other hand, assuming luminous matter is only test particles can only be applied to galaxies with a high mass to light ratio.

The most recent advances considering the interaction of BEC dark matter with luminous matter can be found in [4], where the luminous matter is evolved on top of the gravitational potential due to the BEC dark matter halo of a disk galaxy, but the BEC dark matter is fixed in time. More recently, in [7] two structures collide and the offset of luminous versus dark matter is analyzed; the luminous particles are considered test particles though. In [8] the same problem of galactic encounters is analyzed, but the gravitational interaction among luminous particles is included.

The idea is now to consider all the interactions. The task is non-trivial and some small steps are necessary before addressing the most general scenario. As a step toward including the full interaction between BEC dark matter and luminous matter, in this proceeding we present the construction of an isolated system made of BEC dark matter and luminous matter modeled with N-particles.

Instead of tackling once and for all the general problem of a system made of BEC dark matter and luminous matter, we present a test that the model should pass before being used in the case of many structures interacting among each other. Thus, our result at this point is restricted to a spherically symmetric structure. The structure has two components, the BEC dark matter halo and the luminous matter modeled with N-point particles.

We therefore construct a nearly virialized spherically symmetric structure that can be evolved for a sufficiently long time (millions of years). We include the interaction that the BEC has on the luminous particles, the interaction among the particles themselves and the back reaction of the luminous particles on the BEC gravitational potential.

2. Model of the structure

The BEC halo. The system ruling the BEC is the Gross-Pitaevskii-Poisson coupled system of equations

$$\begin{aligned} i\hbar\frac{\partial\Psi}{\partial t} &= -\frac{\hbar}{2m}\nabla^2\Psi + V\Psi + \frac{2\pi\hbar a}{m^2}|\Psi|^2\Psi, \\ \nabla^2 V &= 4\pi G m|\Psi|^2, \end{aligned} \tag{1}$$

where in general $\Psi = \Psi(t, \mathbf{x})$, m is the mass of the boson, V is the gravitational potential acting as the condensate trap, a is the scattering length of the bosons. This is a coupled system consisting of an evolution equation for Ψ with a potential that is solution of Poisson equation sourced by $|\Psi|^2$.

We choose a BEC dark matter halo to be a ground state equilibrium configuration of the above system (see [9] for a detailed construction), which has the very desirable properties of being stable under radial and axial perturbations [10], and that it is a late-time attractor for the evolution of an initial profile [11]. This configuration has shown to model very well rotation curves of dwarf galaxies with and without a small rotation [5].

The luminous matter. We assume the luminous matter is a system of N-bodies that interact with each other and the gravitational potential due to the BEC halo. These N particles have equal mass M_L/N, where M_L is the mass of the luminous matter of the structure. We spatially distribute the particles such that they satisfy the Plummer density profile

$$\rho(r) = \frac{M_L}{4\pi}\frac{3r_a^2}{(r^2+r_a^2)^{5/2}}, \tag{2}$$

corresponding to the potential $\Phi(r) = \frac{-GM_L}{(r^2+r_a^2)^{1/2}}$. To do that, we start with the cumulative mass distribution

$$m(r) = \int_0^r 4\pi r^2 \rho(r) dr, \tag{3}$$

and using Eq. (2) we integrate for $m(r)$, which gives $m(r) = r^3(r^2+1)^{-3/2}$. Inverting this expression for r we obtain [12]

$$r(m) = (m^{-2/3} - 1)^{-1/2}\,. \tag{4}$$

The procedure to distribute the particles in a three dimensional domain is the following: for every particle we need to find the coordinates r, θ and ϕ of the particle's position. We choose a random number m such that $0 \leq m \leq M_L$ and identify it with the cumulative mass contained within the radius r for that particle. Using Eq (4) we determine the coordinate r for the position of the particle. To determine the angles θ and ϕ we choose two random numbers. The first one between -1 and 1 corresponding to the value of $\cos(\theta)$ and the second one between 0 and 2π corresponding to the value of ϕ.

Now, we proceed to choose the magnitude and direction for the velocity of one of the particles with respect to the center of the distribution. First, the maximum velocity allowed for a particle at a given r is the escape velocity v_e. This velocity can be obtained equating the potential and kinetic energies and solving for v_e to obtain $v_e(r) = \sqrt{2GM_L}(r^2+r_a^2)^{-1/4}$ [12].

Next, the probability $g(v)$ to have a velocity $v = ||\vec{v}||$ at a radial position $r = ||\vec{r}||$ is given by

$$g(v)dv \propto (-E(r,v))^{7/2}v^2 dv\,. \tag{5}$$

In terms of the escape velocity and defining $q = v/v_e$, the distribution function for v becomes:

$$g(q) = (1-q^2)^{7/2}q^2 \text{ with } 0 \leq q \leq 1\,. \tag{6}$$

Instead of inverting Eq.(6) to estimate v, we use a rejection technique. That is, choose a random number x between 0 and 1 and a random number y between 0 and $g_{max} = 0.1$. If $y < g(x)$ we accept the pair (x, y) and compute the velocity as $v = \sqrt{2GM}(x^2 + r_a^2)^{-1/4}$, otherwise, repeat the process choosing another pair (x, y). In this way the particles have randomized peculiar velocities around the center and the configuration is nearly stationary.

3. Tests

The way to know whether the interplay between these two components is correctly implemented is measuring the life-time of the structure. For this, it is necessary to carry out the evolution of the initial configuration and determine the life time of the structure. In Fig. 1 we show various snapshots of an evolved configuration in code units. Each time code unit for a boson mass $m = 10^{-23}eV/c^2$ corresponds to 1.58×10^5yr, (for a detailed translation of all the units for a given boson mass see [5]).

The key advance in this paper is the contribution of the luminous matter to the total gravitational potential due to the BEC. In Fig. 2 we show the gravitational potential of a BEC halo with and without including the luminous matter, for a case with $M_L = 10\%$ of the mass of the BEC configuration and a core radius of luminous matter $r_a = 0.25R_{BEC,95}$, where $R_{BEC,95}$ is the radius that contains 95% of the BEC matter integrated in the whole numerical domain.

A second and more demanding tests is the evolution of the same structure but this time in motion. The structure formation analysis in [1] indicates that high density cored clumps show the same profile as the solitonic solutions of (1), a property that was already shown to happen in [10]. However, one of the most interesting properties of solitonic solutions is that their density does not change with motion and in the most extreme case their profile is retained even after a collision with other soliton.

In the case of the present structure, the matter involved is not only that of the solution of (1) but also involves matter. We want to show how capable our tool is at evolving the initial

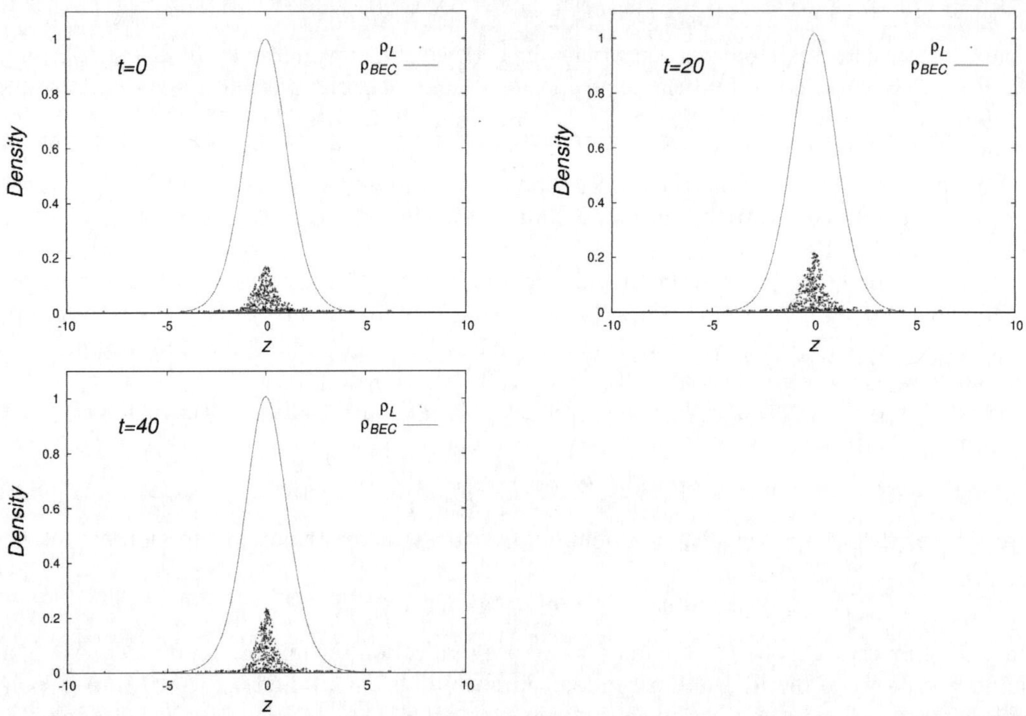

Figure 1. Snapshots of ρ_{BEC} and the distribution of luminous matter ρ_L measured using an SPH method at various times in code units. Both, the wave function and luminous particles are evolving, however remain in a nearly stationary distribution. For $m = 10^{-23}eV/c^2$ the snapshots correspond to 0, 3.16 and 6.32 millions of years.

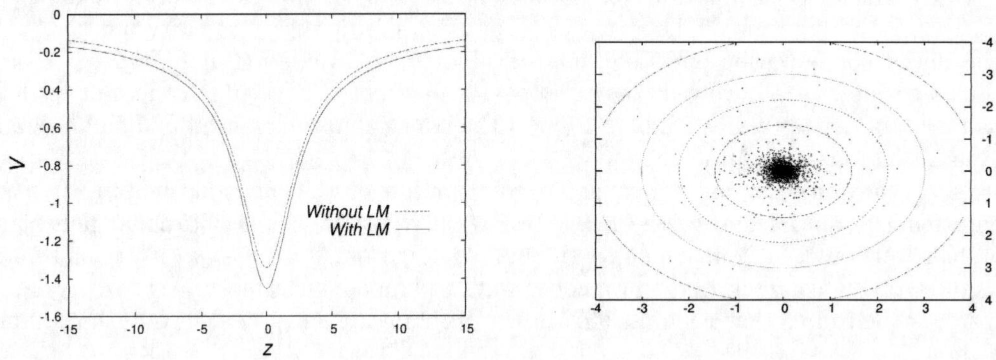

Figure 2. Gravitational potential with and without considering the luminous matter for the parameters of Fig. 1. As expected, the contribution of luminous matter makes the potential well deeper. Also shown is a view from the top of the particles distribution on the xy plane and isocontours of ρ_{BEC}.

configuration when it has received a boost. The boost that is applied to the luminous matter is an addition to the peculiar randomized velocities of the particles of a given velocity, in our case along the z-direction. For the BEC it is slightly different, and the velocity is added as a phase of the wave function Ψ. Both, the phase for the wave function and the velocity of luminous particles in code units is $v_z = 1.5$. For a boson mass $m = 10^{-23} eV/c^2$ the velocity is 2160km/s, which is a very high speed for a galaxy.

In Fig. 3 we show snapshots of the luminous and BEC dark matter, showing the profiles remain nearly time independent, even when the boost is of such magnitude as to expect solitonic behavior according to [10]. In this case, the luminous matter distorts its initial shape and gets more compact. However, after an initial transient the shape stabilizes and evolves following the BEC.

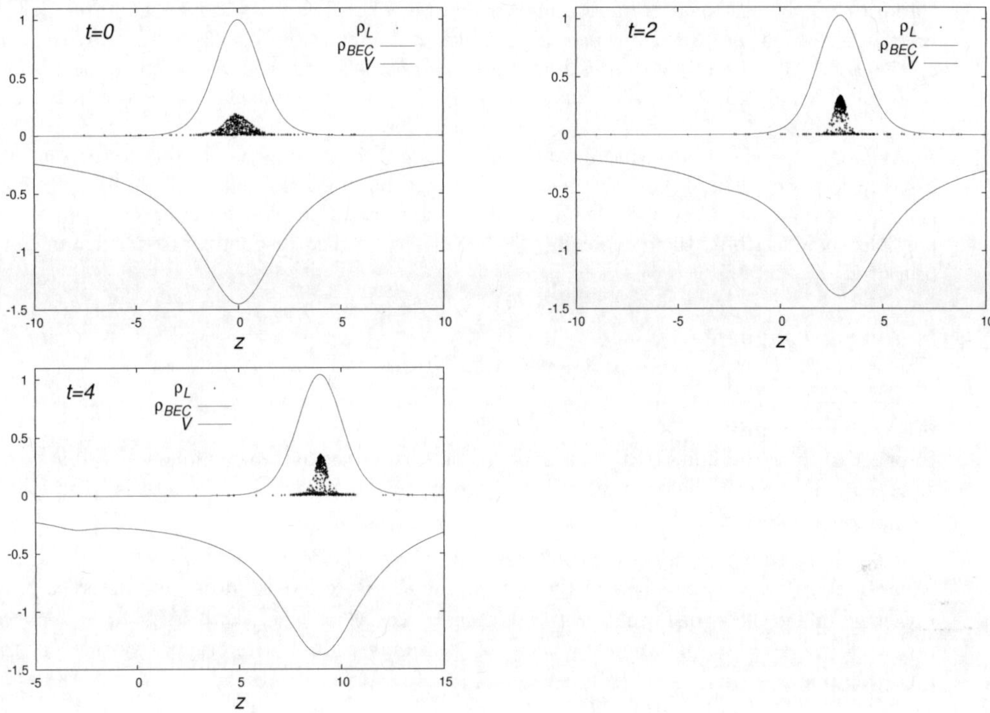

Figure 3. Snapshots of a boosted configuration. In this case the interplay deforms the gravitational potential, unlike the case in which ρ_L does not source Poisson equation in [8].

4. Final comments

We have presented the first BEC dark matter plus luminous matter evolution with full interaction, that is, interaction between the luminous matter, and interaction between the BEC dark matter and luminous matter. Our structure is spherically symmetric but contains all the ingredients for a generalization: randomized peculiar velocities of luminous particles and N-body type of interaction. In a second case, this structure is boosted in order to know whether or not the luminous matter follows the dark matter in a lonely structure scenario. Our result here indicates that luminous matter is a good dark matter tracker, although a more quantitative analysis is required.

This is a small step compared to the advancements in CDM models that include not only luminous particles playing the role of stars, but also include gas, radiation pressure and various other ingredients. Nevertheless, within the BEC dark matter model, this is a significant advance.

This type of initial data in a more general scenario of structure is expected to be very important since it includes now galaxies with a low mass to light ratio, where luminous matter cannot be considered test particles. It will also allow to study the behavior of luminous matter fully coupled during and after the collision of two BEC dark matter structures, which eventually will provide restrictions to the BEC dark matter model.

In order to describe some of the properties of the exampled worked out here, some properties are in turn. The mass of the ground state configuration for the boson mass $m = 10^{-23}eV/c^2$ is of $1.75 \times 10^{10} M_\odot$ with a radius of 3.93kpc, parameters appropriate for a dwarf galaxy. More elaborate configurations of luminous matter and bigger galaxies would involve better BEC halo models, for which there are some possibilities like adding rotation that disperses away the density of the BEC [5] or multi-state configurations that consider various superposed modes of the solutions of the Gross-Pitaevskii equation and show a less compact BEC matter distribution [13].

Aside of the dark matter distribution, our approach is easily generalizable to more arbitrary distributions both, in the density and the velocity field. One would only need to break the isotropy of the case in this paper and use the same selection methods to populate the luminous particles and distribute the components of the velocity field corresponding to -for instance- those of a disk.

5. Acknowledgments

This research is supported by grants CIC-UMSNH 4.9 and 4.23.

6. Appendix: units

In order to show the units treatment and scaling laws of the system we follow [5]. One does not solve eq. (1) directly, before that it is important to remove the constants using the following change of variables $\hat{\Psi} = \frac{\sqrt{4\pi G \hbar}}{mc^2}\Psi$, $\hat{x}_i = \frac{mc}{\hbar}x_i$ (for cartesian coordinates x_i), $\hat{t} = \frac{mc^2}{\hbar}t$, $\hat{V} = \frac{V}{mc^2}$, $\hat{a} \rightarrow \frac{c^2}{2mG}a$, so that the numerical coefficients $\hbar$, $\hbar/m$, $2\pi\hbar/m^2$, $4\pi Gm$ do not appear in (1). Thus by fixing the boson mass m the hatted variables are fixed. Moreover, the system (1) is invariant under the transformation $t = \lambda^2\hat{t}$, $x_i = \lambda\hat{x}_i$, $\Psi = \hat{\Psi}/\lambda^2$, $V = \hat{V}/\lambda^2$, $a = \lambda^2\hat{a}$, for an arbitrary value of the parameter λ [9]. This rescaling reduces the original system (1) to the following one

$$
\begin{aligned}
i\frac{\partial\Psi}{\partial t} &= -\frac{1}{2}\nabla^2\Psi + V\Psi + a|\Psi|^2\Psi \qquad (7)\\
\nabla^2 V &= |\Psi|^2, \qquad (8)
\end{aligned}
$$

which is the one we solve numerically and is written in what we call code units. If one calculates one solution for the non-hatted variables, say for a given λ, other solutions are automatically found for any other value of λ, that is, a new solution for the hatted and tilde variables is found just by using a new value of this parameter.

The code and physical units are related once the scale invariance parameter λ is fixed. This can be achieved by fixing for instance the mass of the configuration to the value of a given galaxy mass, or by choosing a spatial scale in given units. Here we choose the second approach following [5]. Specifically, since $\lambda = \frac{\hbar}{mc}\frac{x_i}{\tilde{x}_i}$ (or equivalently the spherical coordinate r), if physical measurements of space $\tilde{x}_i$ are given in kpc, we choose the code coordinates x_i to represent kpc as well. Thus it suffices to write the factor $\frac{\hbar}{mc}$ in kpc. For $m = 10^{-23}\mathrm{eV}/c^2$ its value is $\lambda = \frac{\hbar}{mc}[kpc]\frac{x}{\bar{x}[kpc]} = 0.0006389$.

In this paper we use only one value of the boson mass. If one chooses another value of the boson mass, say $m = 10^{-22}\mathrm{eV}/c^2$ as suggested in [14], the effect would be that $\lambda = \frac{\hbar}{mc}[kpc]\frac{x}{\bar{x}[kpc]} = 0.006389$. With this, the spatial and time scales would change according to the scaling relations described above.

References

[1] H-Y Schive, T. Chiueh and T. Broadhurst, Nat. Phys. 10, 496 (2014).
[2] T. Harko, JCAP 1105, 22 (2011).
[3] D. J. E. Marsh and A-R. Pop, MNRAS, 451, 2479 (2015).
[4] L. A. Martínez-Medina and T. Matos, MNRAS 444, 185-191 (2014).
[5] F. S. Guzmán and F. D. Lora-Clavijo, Gen. Rel. Grav. 47, 21 (2015).
[6] S-J. Sin, Phys. Rev. D 50, 3650 (1994). S.U. Ji and S-J. Sin, Phys. Rev. D 50, 3655 (1994). F. E. Schunk, astro-ph/9802258. F. S. Guzmán, T. Matos and H. Villegas-Brena, Astron. Nachr. 320, 97 (1999). F. S. Guzmán and T. Matos, Class. Quantum Grav., 17, L9-L16 (2000). A. Arbey, J. Lesgourgues, P. Salati, Phys. Rev. D 64, 123528 (2001). A. Arbey, J. Lesgourgues, P. Salati, Phys. Rev. D 68, 023511 (2003). C. G. Boehmer and T. Harko, JCAP 0706, 025 (2007). L. A. Ureña-López and A. Bernal, Phys. Rev. D 82, 123535 (2010). F. S. Guzmán and F. D. Lora-Clavijo, Gen. Rel. Grav. 47, 21 (2015).
[7] A. Paredes and H. Michinel, Physics of the Dark Universe 12, 5055 (2016).
[8] F. S. Guzmán, J. A. González, J. P. Cruz, Phys. Rev. D 93, 103535 (2016).
[9] F. S. Guzmán and L. A. Ureña-López, Phys. Rev. D 69, 124033 (2004).
[10] A. Bernal and F. S. Guzmán, Phys. Rev. D 74, 103002 (2006).
[11] F. S. Guzmán and L. A. Ureña-López, ApJ 645, 814-819 (2006).
[12] S. J. Aarseth, M. Hénon and R. Wielen, A & A 37, 183-187 (1974). P. Hut and J. Makino, The Art of Computational Science: The Kali Code.
[13] L. A. Ureña-López and A. Bernal, Phys. Rev. D 82, 123535 (2010).
[14] S-R Chen, H-Y Schive, T. Chiueh, arXiv:1606.09030 [astro-ph.GA]

IOP Conf. Series: Journal of Physics: Conf. Series **1010** (2018) 012005 doi:10.1088/1742-6596/1010/1/012005

Comparison between two scalar field models using rotation curves of spiral galaxies

Lizbeth M. Fernández-Hernández[1], **Mario A. Rodríguez-Meza**[1,3] **and Tonatiuh Matos**[2]

[1]Departamento de Física, Instituto Nacional de Investigaciones Nucleares, Apdo. Postal 18-1027, México D.F. 11801, México

[2]Departamento de Física, Centro de Investigación y de Estudios Avanzados del IPN, A.P. 14-740, 07000 México D. F., México

E-mail: [3]`marioalberto.rodriguez@inin.gob.mx (Corresponding author)`

Abstract. Scalar fields have been used as candidates for dark matter in the universe, from axions with masses $\sim 10^{-5}$eV until ultra-light scalar fields with masses $\sim 10^{-22}$eV. Axions behave as cold dark matter while the ultra-light scalar fields galaxies are Bose-Einstein condensate drops. The ultra-light scalar fields are also called scalar field dark matter model. In this work we study rotation curves for low surface brightness spiral galaxies using two scalar field models: the Gross-Pitaevskii Bose-Einstein condensate in the Thomas-Fermi approximation and a scalar field solution of the Klein-Gordon equation. We also used the zero disk approximation galaxy model where photometric data is not considered, only the scalar field dark matter model contribution to rotation curve is taken into account. From the best-fitting analysis of the galaxy catalog we use, we found the range of values of the fitting parameters: the length scale and the central density. The worst fitting results (values of χ^2_{red} much greater than 1, on the average) were for the Thomas-Fermi models, i.e., the scalar field dark matter is better than the Thomas-Fermi approximation model to fit the rotation curves of the analysed galaxies. To complete our analysis we compute from the fitting parameters the mass of the scalar field models and two astrophysical quantities of interest, the dynamical dark matter mass within 300 pc and the characteristic central surface density of the dark matter models. We found that the value of the central mass within 300 pc is in agreement with previous reported results, that this mass is $\approx 10^7\ M_\odot/\text{pc}^2$, independent of the dark matter model. And, on the contrary, the value of the characteristic central surface density do depend on the dark matter model.

1. Introduction

When anisotropies were discovered in the cosmic microwave background (CMB) by the Cosmic Background Explorer (COBE) satellite a standard model for the universe was almost stablished. Planck Mission is the third generation space mission, following COBE and the Wilkinson Microwave Anisotropy Probe (WMAP), that measured the CMB anisotropies. Planck collaboration updated the values for the cosmological parameters, temperature and polarization anisotropies of the CMB radiation, showing that these observations are consistent with the standard spatially-flat six-parameters model of the universe: cosmological constant (Λ) and cold dark matter (CDM), the so called ΛCDM cosmology[1].

With the study of the rotation curve (RC) of spiral galaxies [2], the existence of dark matter is gained observational support, giving rise to several models to study this "extra" component of matter, not visible, that modifies the Keplerian curve, see for example [2, 3, 4]. This "extra" matter component known as dark matter (DM) is the difference between the mass of the galaxy predicted by its luminosity and its mass predicted by the rotation velocities of the stars and gas. The study of RC gives one of the strongest evidences that spiral galaxies are embedded in extended halos of DM.

When the evolution of the universe is studied with more detail, several models have been proposed to explain this evolution. It turns out that the most accepted model in cosmology is the ΛCDM, which uses the concept of CDM, a kind of non baryonic and non relativistic particle at the decoupling epoch. However, it has several problems at galactic scales and less. This is the reason why a great amount of models have arisen as alternative models to explain the DM of galaxies. One of them is the scalar field dark matter model, determined by a fundamental scalar field φ [5], with a minimal coupling with the metric, or a scalar field coupled non minimally with the metric that have been used to study large scale structure formation [6] or galactic dynamics [7, 8, 9]. Recently, numerical simulations of large scale structure formation using a scalar field dark matter has given a strong support to these kind of dark matter models at galactic scales [10].

Scalar field based dark matter models come in different flavours depending on their equations of motion [5, 6, 11, 12, 13, 14]. In this work we want to test two proposals using a fitting analysis of the rotation curves of LSB galaxies (dark matter dominated galaxies, ideal objects to test different dark matter models). In particular, in this paper we are going to study three scalar field based models to explain DM: a Gross-Pitaevskii Bose-Einstein condensate in the Thomas-Fermi approximation (TF), with [15] and without [16] cut at a characteristic radial distance, and a scalar field solution of the Klein-Gordon equation, that we simply refer to as scalar field dark matter model (SFDM) [17]. The TF models have problems to fit the RC because the mass density profile may take negative values beyond the characteristic radial distance when we do not use a cut in the density. This problem does not appear in the SFDM model where the density profile is always positive. The fitting analysis will test the models and provide the parameters that characterize the dark matter model density profiles, one of them is the mass of the scalar field. In order to do so, this work is organized as follows. In section 2 we briefly explain the two scalar field models and the different approaches to be compared, the Thomas-Fermi approximation and the scalar field dark matter. In section 3 we make the comparison between the TF and SFDM models and we find the best fitting parameters using a LSB galaxy RC catalog. In section 4 we give our conclusions of this comparison.

2. The scalar field models

For the SFDM model we use the density profile as given by [17]

$$\rho_{SFDM}(r) = \rho_s \frac{\sin^2(r/r_s)}{(r/r_s)^2} \, . \tag{1}$$

known as the flat oscillaton. This corresponds to the static solution of the corresponding Klein-Gordon equation with a quadratic potential for the scalar field φ, $V(\varphi) = m_\varphi^2 \varphi^2/2$, m_φ is the scalar field mass and is given by the inverse of r_s through the Compton relationship $r_s = \hbar c/m_\varphi$, $\hbar$ is the Planck's constant and c the speed of light. See Ref. [17] for details.

The corresponding rotation velocity is

$$V^2_{SFDM}(r) = -\frac{\rho_s}{4} r_s^2 \left(-2 + \frac{r_s}{r} \frac{\sin(2r/r_s)}{r/r_s} \right) . \tag{2}$$

The other SF model we want to consider is the scalar field that satisfies the Gross-Pitaevskii and Poisson equations. These two equations have, in the Thomas-Fermi approximation, the solution [18]

$$\rho_{TF}(r) = \rho_s \frac{\sin(kr)}{kr} \,, \tag{3}$$

where ρ_s is the central density of the condensate, $\rho_s = \rho(0)$. The total mass of the galactic dark matter halo is

$$M_{TF}(r) = \frac{4\pi\rho r}{k^2}\left(1 - kr\cot(kr)\right) . \tag{4}$$

This expression has a relevant significance only when the mass distribution is inside of a boundary surface with radius R (Thomas-Fermi approximation with a cut), where $\rho(R) = 0$ and then $kR = \pi$

$$R = \pi\sqrt{\frac{\hbar^2 a}{G m_\psi^3}} \,, \tag{5}$$

where G is the Newton gravitational constant and a and m_ψ are the scattering length and the mass of the scalar field particle, respectively. This expression fix the radius of the condensate dark matter halo. We follow [18] and chose $a = 1$ fm. Hence, the tangential velocity is

$$V_{TF}^2(r) = \frac{4G\rho_s R^2}{\pi}\left[\frac{\sin(\pi r/R)}{\pi r/R} - \cos\left(\frac{\pi r}{R}\right)\right] , \tag{6}$$

for $r > R$ the rotation curves follow the standard Keplerian law.

If we considered that the previous results are valid even outside of the boundary surface (Thomas-Fermi approximation without a cut [16]), we have two scalar field models in the TF approximation. In this case we will have that equation (6) is without constriction.

Therefore, for spherical symmetry, our SF galaxy model will be as follows. A test particle will move under the action of the potential $\Phi_{DM}(r)$ which is the potential due to a DM mass distribution, where

$$\Phi_{DM}(r) = \begin{cases} \Phi_{TF} & = \begin{cases} \text{with cut} \\ \text{without cut} \end{cases} \\ \\ \Phi_{SFDM} \end{cases} \tag{7}$$

The potential of the SF, $\Phi_{DM}(r)$ is the solution the corresponding Klein-Gordon equation for the SFDM model [17] or it is the solution for Gross-Pitaevskii and Poisson equations in the Thomas-Fermi approximation [18].

The rotational velocities as given in Eqs. (2) and (6) (with and without a cut) were obtained using

$$V^2 = r\left|\frac{d\Phi_{DM}}{dr}\right| . \tag{8}$$

3. Results

In this section we perform the comparison among models in Eq. (7) that give the theoretical RC that we test against observed RC by a fitting analysis.

We are using a high resolution LSB galaxies catalog reported in [19]. LSB galaxies are the ideal objects to test the distribution of dark matter in galaxies and to compare different dark matter models. The authors in Ref. [19] considered the visible data contribution for classifying the galaxies according to availability of photometric data and use three models: zero disk model, where the dark matter is the main component in the halos, making zero all the visible components of the galaxies; galaxies with photometry with a constant ratio M/L_* and the maximum disk model. In our case only galaxies without photometry data are considered [20] and we will use the zero disk model.

Therefore in this work we use only the zero disk model where the velocity contributions from baryons are ignored, they are test particles, and the total RC has only the contribution of the dark matter models. Accordingly we consider only galaxies classified as without photometric data in Ref. [19]. These galaxies belong to the sample for which an optical rotation curve was possible to measure but there were no optical nor H I photometry, implying that mass models for the stars or gas were not possible to build.

We perform a minimum χ^2 analysis to fit the observations with $\chi^2 = \sum_{i=1}\left(\frac{V_i - V(r_i)}{\sigma_i}\right)^2$, where V_i is the observed velocity, σ_i is the error in the measurement of V_i and $V(r_i)$ is the theoretical value of the velocity, which is computed at the same position where V_i was measured. For $V(r_i)$ we will use each one of the models in Eq. (7). We use units where the length is in kpc, the velocity in km/s and $G = 1$. The free parameters to fit are the following: for the SFDM model, the RC equation is given by Eq. (2), it has two parameters, ρ_s and r_s; for the TF models the RC is given by Eq. (6) then it has also two parameters, ρ_s and R. In the fitting analysis we impose that all these parameters are always positive and without an upper limit. To complete the data analysis we compute the errors in the fitting parameters and the Q-value that gives us the goodness of fit [21].

In Fig. (1) we show the comparison between the observed RC and the theoretical RC as given by fitting each of the DM models for the galaxies without photometry. The left column show the TF model without cut, the middle column the TF model with cut and the right column SFDM flat oscillaton model. It is very clear that the TF approximation without cut is the worst case, as can be seen, for instance, for galaxies ESO 0140040 and UGC 11748 whereas the best fitting results were obtained for flat oscillaton model as can be see in the column on the right. This is confirmed in Tables 1–3 (see below) in columns (9) where are shown the Q-values, for the TF without cut model they are very small or zero for a lot of galaxies.

Table 1 shows the best fitting parameters for galaxies without photometry for the scalar field dark matter model in the TF approximation without cut, using Eq. (6). Fitting parameters are shown in columns (2) and (3). From column (3) and using Eq. (5) (we have chosen $a = 1$ fm according to Ref. [18]) we obtain column (4), the mass of the scalar field particle, its values are in the range of 9.86 meV $< m_\psi <$ 51.11 meV, consistent with the upper limit $m_\psi < 1.87$ eV computed from cosmological considerations [22]. Fitting parameters are consistent with values found in Ref. [16] except for galaxies, UGC 11616, UGC 11748 and UGC 11819, due to that in Ref. [16] some measured values are missing. This DM model has the problem that density can take negative values and so the rotation velocity (leading to instabilities) as can be noticed clearly in Fig. (1) left column, galaxies ESO 0140040 and UGC 11748. In column (5) we show the characteristic surface density, $\mu_{DM} = \rho_s r_c$, r_c is the radius at which the density has decayed to a half of the central density value, ρ_s. Its mean value is, 225.7 $M_\odot/\mathrm{pc}^2$. In column (6) it is shown the dynamical mass of the dark matter up to 300 pc. Its mean value is 0.6×10^7 $M_\odot$ in agreement with the value found for dSph Milky Way satellites [23]. And in column (7) we show the total mass of this DM model.

We make the same analysis for Table 2 which present the results for the TF approximation with a cut. The scalar field mass is in the range, 10.43 meV $< m_\psi <$ 51.11 meV, almost the same as the TF without a cut. The mean value of the characteristic surface density is 244.4 $M_\odot/\mathrm{pc}^2$ and the mean value of the mass within 300 pc is 0.7×10^7 $M_\odot$.

For the SFDM model we can compute the same statistical information as was presented above, the results are shown in Table 3. In this case the characteristic surface density is $\mu_{DM} = \rho_s r_s$, r_s is the scale radius and ρ_s is the central density value of the flat oscillaton. Its mean value is 186.7 $M_\odot/\mathrm{pc}^2$. The DM mass within 300 pc mean value is 0.9×10^7 $M_\odot$ that is also in agreement with the result found in Ref. [23]. The range of the oscillaton mass is 1×10^{-27} eV $< m_\varphi <$ 10.45×10^{-27} eV. The mass of the flat oscillaton was computed using the Compton

IOP Conf. Series: Journal of Physics: Conf. Series **1010** (2018) 012005 doi:10.1088/1742-6596/1010/1/012005

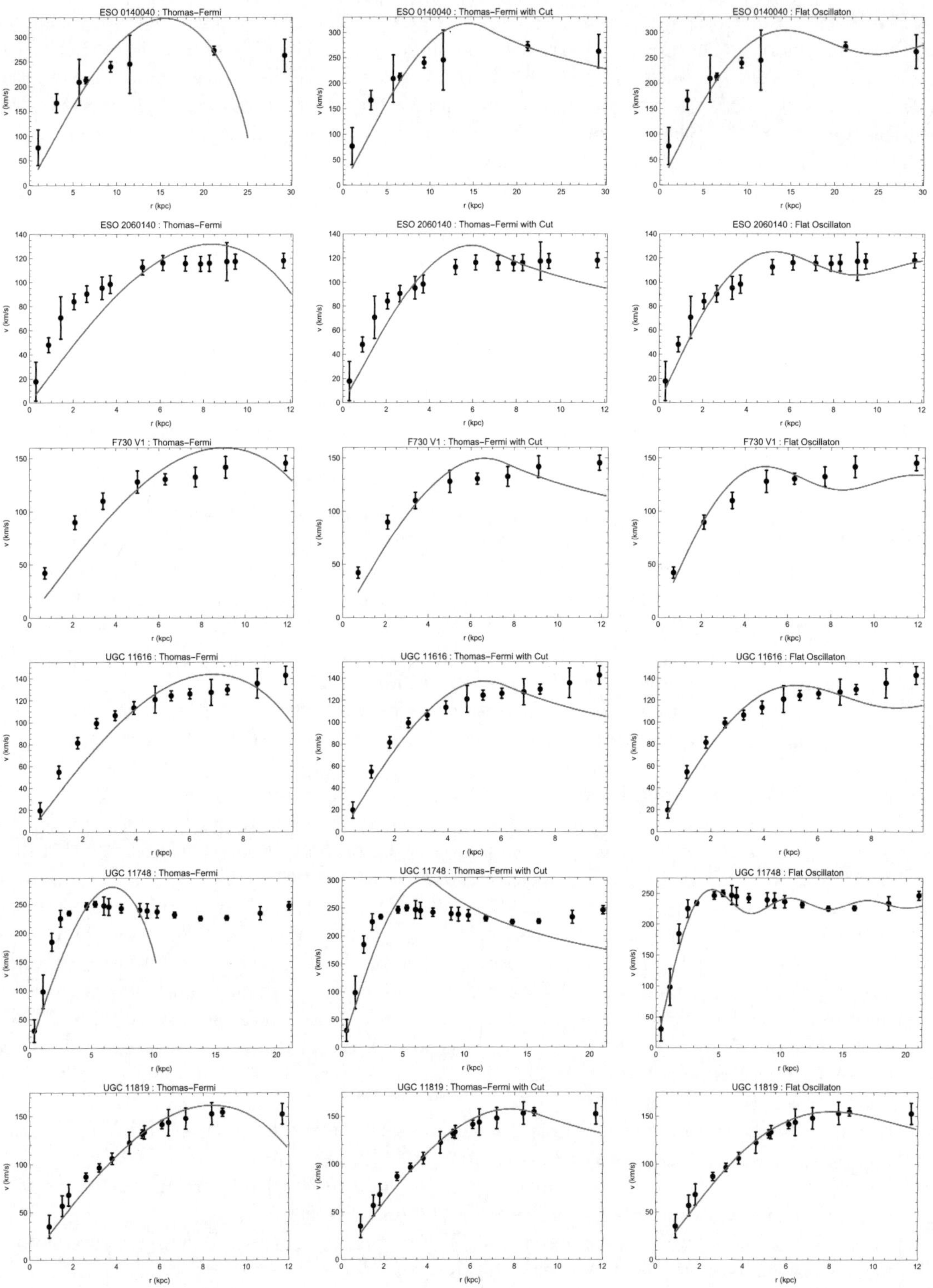

Figure 1. LSB galaxies fit with the zero disk and the scalar field models, Eq. (7). Black solid circles are the observed RC. The blue line is the theoretical RC obtained by the fitting method.

IOP Conf. Series: Journal of Physics: Conf. Series **1010** (2018) 012005 doi:10.1088/1742-6596/1010/1/012005

Galaxy sample Thomas-Fermi DM Model Best Fitting Parameters								
Galaxy	ρ_s ($10^{-3} M_\odot/\mathrm{pc}^3$)	R (kpc)	m_ψ (meV)	μ_{DM} ($M_\odot/\mathrm{pc}^2$)	M_{DM}(300 pc) ($10^7 M_\odot$)	$M_{DM}(r_{\max})$ ($10^{11} M_\odot$)	χ^2_{red}	Q-value
(1)	(2)	(3)	(4)	(5)	(6)	(7)	(8)	(9)
ESO 0140040	61.71±4.98	17.84±0.37	9.86	664.20	0.70	5.25	4.15	0.0004
ESO 0840411	4.69±1.35	11.82±3.61	12.98	33.43	0.05	0.08	0.10	0.9982
ESO 1200211	13.66±3.84	2.92±0.41	32.95	24.07	0.15	0.003	0.44	0.9476
ESO 1870510	32.55±7.35	2.93±0.38	32.85	57.63	0.36	0.01	0.21	0.9932
ESO 2060140	32.78±2.67	9.54±0.29	14.97	188.69	0.37	0.23	8.43	0.
ESO 3020120	22.74±4.70	8.86±0.60	15.73	121.50	0.26	0.10	0.92	0.5046
ESO 3050090	21.50±6.61	4.81±0.99	23.62	62.43	0.24	0.03	0.10	1.000
ESO 4250180	13.53±4.44	15.63±4.21	10.77	127.58	0.15	0.65	0.33	0.9237
ESO 4880049	54.29±9.05	5.36±0.45	21.98	175.72	0.61	0.09	0.80	0.6176
F730 V1	41.37±4.02	10.31±0.41	14.22	257.24	0.47	0.47	11.62	0.
UGC 4115	142.22±45.79	1.53±0.82	50.77	131.09	1.55	0.004	0.007	1.000
UGC 11454	41.69±2.22	10.43±0.24	14.11	262.25	0.47	0.51	12.91	0.
UGC 11557	14.18±4.10	9.39±4.12	15.13	80.34	0.16	0.09	0.08	0.9997
UGC 11583	89.60±28.62	1.51±0.29	51.11	81.77	0.97	0.004	0.07	0.9998
UGC 11616	58.13±3.76	7.83±0.24	17.08	274.51	0.66	0.23	10.09	0.
UGC 11648	25.90±1.31	11.56±0.27	13.17	180.59	0.29	0.47	15.63	0.
UGC 11748	228.42±5.52	7.67±0.01	17.32	1056.49	2.58	2.96	58.47	0.
UGC 11819	48.62±2.84	9.63±0.33	14.88	282.48	0.55	0.39	2.11	0.0109

Table 1. We show the fitting parameters, derived quantities (see text for details) and the χ^2_{red} and Q-value for galaxies without photometry, for the Thomas-Fermi model without cut. Column (1) is the name of the galaxy; columns (2) and (3) give the fitting parameters; columns (4)–(7) give derived quantities and columns (8) and (9) give the χ^2_{red} and Q-value, respectively.

wave length relationship [17], r_s shown in column (3) of Table 3. This range of values for the flat oscillaton rule out this model according to the results presented in Refs. [11, 14].

In column (7) of Tables 1–3 we show the estimated values of the DM models total mass. Their mean values are: for the TF model without cut, $6.4 \times 10^{10}\ M_\odot$, for the TF model with a cut, $4.7 \times 10^{10}\ M_\odot$ and $6.5 \times 10^{10}\ M_\odot$ for the oscillaton model. All these value are very similar and are consistent for the values of LSB galaxies.

4. Conclusions

In this paper we have analysed three scalar field DM models using a RC LSB galaxies catalog reported in [19]. We found a better χ^2_{red} and Q-values when using the SFDM model versus TF models. TF approximation without cut presents problems to fit the RC because the profile density function, $\sin kr/kr$, has a strong decay when R has the smallest values, giving negative values for the mass density, leading to instabilities of the circular orbits. In general TF approximation with and without a cut has problems to fit the observational data.

For each model and for each galaxy in the catalog we computed two important quantities, the surface density μ_{DM} and the mass within 300 pc, M_{DM}(300 pc) for the DM haloes. The values

Galaxy sample Thomas-Fermi with Cut DM Model Best Fitting Parameters								
Galaxy	ρ_s $(10^{-3}M_\odot/\text{pc}^3)$	R (kpc)	m_ψ (meV)	μ_{DM} $(M_\odot/\text{pc}^2)$	$M_{DM}(300\text{ pc})$ $(10^7 M_\odot)$	$M_{DM}(r_{\max})$ $(10^{11}M_\odot)$	χ^2_{red}	Q-value
(1)	(2)	(3)	(4)	(5)	(6)	(7)	(8)	(9)
ESO 0140040	64.38±4.22	16.40±0.18	10.43	637.04	0.73	3.62	3.87	0.0007
ESO 0840411	4.69±1.35	11.82±3.61	12.98	33.43	0.05	0.08	0.10	0.9982
ESO 1200211	19.57±4.08	2.29±0.06	38.77	27.03	0.22	0.003	0.29	0.9913
ESO 1870510	32.82±7.28	2.91±0.37	33.01	57.70	0.37	0.01	0.21	0.9935
ESO 2060140	63.56±3.97	6.78±0.04	18.80	259.98	0.72	0.25	3.83	3.3×10^{-6}
ESO 3020120	29.46±5.25	7.24±0.16	18.00	128.65	0.33	0.14	0.42	0.9245
ESO 3050090	21.50±6.61	4.81±0.99	23.62	62.43	0.24	0.03	0.10	1.000
ESO 4250180	13.53±4.44	15.63±4.21	10.77	127.58	0.15	0.65	0.33	0.9237
ESO 4880049	54.90±8.95	5.33±0.43	22.08	176.44	0.62	0.11	0.78	0.6356
F730 V1	66.70±6.19	7.58±0.1	17.45	305.14	0.75	0.37	9.01	7.2×10^{-10}
UGC 4115	142.22±45.79	1.53±0.82	50.77	131.09	1.55	0.004	0.007	1.000
UGC 11454	47.99±2.00	9.47±0.12	15.05	274.12	0.54	0.52	10.91	0.
UGC 11557	14.18±4.10	9.39±4.12	15.13	80.34	0.16	0.09	0.08	0.9997
UGC 11583	89.60±28.62	1.51±0.29	51.11	81.77	0.97	0.004	0.07	0.9998
UGC 11616	86.20±4.68	6.13±0.07	20.11	318.68	0.97	0.25	5.29	2.1×10^{-8}
UGC 11648	28.09±1.22	10.96±0.18	13.65	185.74	0.32	0.47	15.03	0.
UGC 11748	264.30±6.15	7.69±0.01	17.29	1225.92	2.98	1.53	24.22	0.
UGC 11819	52.90±2.62	8.98±0.19	15.59	286.44	0.60	0.49	1.19	0.2789

Table 2. We show the fitting parameters, derived quantities, the χ^2_{red} and the Q-value for the LSB galaxies for the Thomas-Fermi model with cut. The meaning of columns is the same as in Table 1.

we obtained are roughly constant and independent of the absolute magnitude of the galaxies as was also found in Ref. [24]. The mean value for $M_{DM}(300\text{ pc})$ is consistent with the common mass for dwarf spheroidal galaxies of the order of 10^7 $M_\odot$ reported in [23] and independent of the DM model. On the contrary, the mean values of μ_{DM} do depend of the DM model, 225.7 and 244.4 $M_\odot/\text{pc}^2$ for TF models versus 186.7 $M_\odot/\text{pc}^2$ for the flat oscillaton.

Galaxy sample Flat Oscillaton DM Model Best Fitting Parameters								
Galaxy	ρ_s $(10^{-3}M_\odot/\text{pc}^3)$	r_s (kpc)	m_φ $(10^{-27}$ eV)	μ_{DM} $(M_\odot/\text{pc}^2)$	$M_{DM}(300$ pc) $(10^7 M_\odot)$	$M_{DM}(r_{\max})$ $(10^{11}M_\odot)$	χ^2_{red}	Q-value
(1)	(2)	(3)	(4)	(5)	(6)	(7)	(8)	(9)
(1) ESO 0140040	68.95±6.67	6.39±0.35	1.00	440.40	0.78	5.28	3.14	0.005
(2) ESO 0840411	4.78±1.47	5.03±1.70	1.27	24.05	0.05	0.08	0.09	1.0
(3) ESO 1200211	33.68±26.22	0.70±0.28	9.10	23.70	0.37	0.004	0.17	1.0
(4) ESO 1870510	35.31±9.45	1.16±0.21	5.52	40.96	0.39	0.01	0.15	1.0
(5) ESO 2060140	87.65±18.27	2.33±0.25	2.75	204.29	0.99	0.39	1.96	0.02
(6) ESO 3020120	34.19±14.29	2.68±0.79	2.39	91.60	0.39	0.16	0.25	0.99
(7) ESO 3050090	22.41±7.53	1.98±0.49	3.23	44.45	0.25	0.03	0.09	1.0
(8) ESO 4250180	14.90±6.45	6.19±2.15	1.03	92.27	0.17	0.65	0.28	0.95
(9) ESO 4880049	61.12±12.50	2.09±0.24	3.07	127.48	0.69	0.11	0.56	0.83
(10) F730 V1	126.25±22.69	2.20±0.19	2.91	277.74	1.42	0.51	3.03	0.006
(11) UGC 4115	143.25±48.08	0.66±0.38	9.63	95.15	1.56	0.004	0.007	1.0
(12) UGC 11454	53.91±4.61	3.68±0.17	1.74	198.11	0.61	0.53	8.48	0.
(13) UGC 11557	14.36±4.39	4.05±1.91	1.58	58.10	0.16	0.09	0.07	1.0
(14) UGC 11583	95.54±35.59	0.61±0.15	10.45	58.49	1.03	0.004	0.06	1.0
(15) UGC 11616	101.60±13.32	2.31±0.16	2.77	234.48	1.15	0.30	3.46	0.0001
(16) UGC 11648	31.12±2.06	4.27±0.16	1.50	132.94	0.35	0.48	13.14	0.
(17) UGC 11748	514.54±55.02	1.97±0.11	3.25	1012.67	5.79	2.60	2.19	0.004
(18) UGC 11819	56.70±4.49	3.59±0.19	1.78	203.35	0.64	0.52	0.76	0.71

Table 3. In this table we show the fitting parameters, derived quantities, the χ^2_{red} and the Q-value for the analysed LSB galaxies for the flat oscillaton DM model.

References

[1] Planck Collaboration, Ade P A R, Aghanim N, Arnaud M, Ashdown M, Aumont J, Baccigalupi C, Banday A J, Barreiro R B, Bartlett J G and et al 2015 *ArXiv e-prints* (*Preprint* 1502.01589)

[2] Sofue Y and Rubin V 2001 *Ann. Rev. Astron. Astrophys.* **39** 137–174 (*Preprint* astro-ph/0010594)

[3] Begeman K G, Broeils A H and Sanders R H 1991 *Mon. Not. Roy. Astron. Soc.* **249** 523

[4] Burkert A 1996 *IAU Symp.* **171** 175 [Astrophys. J.447,L25(1995)] (*Preprint* astro-ph/9504041)

[5] Matos T, Vázquez-González A and Magaña J 2008 *American Institute of Physics Conference Series* (*American Institute of Physics Conference Series* vol 1083) ed Herrera-Aguilar A, Guzmán-Murillo F S, Nucamendi-Gómez U and Quiros I pp 144–170

[6] Rodríguez-Meza M A 2012 *Advances in Astronomy* **2012** 509682 (*Preprint* 1112.5201)

[7] Rodriguez-Meza M A, Klapp J, Cervantes-Cota J L and Dehnen H 2001 *The influence of scalar fields in protogalactic interactions* (Kluwer Academic Publishers) p 213 (*Preprint* 0907.2466)

[8] Cervantes-Cota J L, Rodríguez-Meza M A, Gabbasov R and Klapp J 2007 *Revista Mexicana de Fisica Supplement* **53** 22–32 (*Preprint* 0704.2579)

[9] Rodríguez-Meza M A 2012 *American Institute of Physics Conference Series* (*American Institute of Physics Conference Series* vol 1473) ed Ureña-López L A, Becerril-Bárcenas R and Linares-Romero R pp 74–84 (*Preprint* 1210.6947)

[10] Schive H Y, Chiueh T and Broadhurst T 2014 *Nature Phys.* **10** 496–499 (*Preprint* 1406.6586)

[11] Rodríguez-Meza M A, Hernández-Almada A and Matos T 2012 *Revista Mexicana de Fisica E* **58** 53–60

[12] Hernández-Almada A, Rodríguez-Meza M A and Matos T 2011 *American Institute of Physics Conference Series* (*American Institute of Physics Conference Series* vol 1396) ed Ureña-López L A, Aurelio Morales-Técotl H, Linares-Romero R, Santos-Rodríguez E and Estrada-Jiménez S pp 196–201

[13] Matos T and Rodríguez-Meza M A 2014 *Journal of Physics Conference Series* (*Journal of Physics Conference Series* vol 545) p 012009 (*Preprint* 1405.6649)

IOP Conf. Series: Journal of Physics: Conf. Series **1010** (2018) 012005 doi:10.1088/1742-6596/1010/1/012005

[14] Bernal T, Fernández-Hernández L M, Matos T and Rodríguez-Meza M A 2017 *ArXiv e-prints* (*Preprint* 1701.00912)
[15] Diez-Tejedor A, Gonzalez-Morales A X and Profumo S 2014 *Phys. Rev.* **D90** 043517 (*Preprint* 1404.1054)
[16] Robles V H and Matos T 2012 *Mon. Not. Roy. Astron. Soc.* **422** 282–289 (*Preprint* 1201.3032)
[17] Ureña López L A, Matos T and Becerril R 2002 *Class. Quant. Grav.* **19** 6259–6277
[18] Harko T 2011 *JCAP* **1105** 022 (*Preprint* 1105.2996)
[19] de Blok W J G, McGaugh S S and Rubin V C 2001 *Astron. J.* **122** 2396–2427
[20] Kuzio de Naray R, McGaugh S S and de Blok W J G 2008 *Astrophys. J.* **676** 920–943 (*Preprint* 0712.0860)
[21] Press W, Teukolsky S A, Vetterling W T and Flannery B P 2007 *Numerical Recipes* (Cambridge University Press.)
[22] Boyanovsky D, de Vega H J and Sanchez N G 2008 *Phys. Rev. D* **77** 043518 (*Preprint* 0710.5180)
[23] Strigari L E, Bullock J S, Kaplinghat M, Simon J D, Geha M, Willman B and Walker M G 2008 *Nature* **454** 1096–1097 (*Preprint* 0808.3772)
[24] Garcia-Aspeitia M A, Lopez-Dominguez J C, Ortiz C, Hinojosa-Ruiz S and Rodriguez-Meza M A 2017 *Rev. Mex. Fis.* **63** 423–438 (*Preprint* 1511.06740)

Constraining brane tension using rotation curves of galaxies

Miguel A. García-Aspeitia[1,2] **and Mario A. Rodríguez-Meza**[3,4]

[1]Consejo Nacional de Ciencia y Tecnología,
Av. Insurgentes Sur 1582. Colonia Crédito Constructor, Del. Benito Juárez C.P. 03940, México D.F. México

[2]Unidad Académica de Física, Universidad Autónoma de Zacatecas, Calzada Solidaridad esquina con Paseo a la Bufa S/N C.P. 98060, Zacatecas, México

[3]Departamento de Física, Instituto Nacional de Investigaciones Nucleares, Apdo. Postal 18-1027, México D.F. 11801, México

E-mail: [4]`marioalberto.rodriguez@inin.gob.mx (Corresponding author)`

Abstract. We present in this work a study of brane theory phenomenology focusing on the brane tension parameter, which is the main observable of the theory. We show the modifications steaming from the presence of branes in the rotation curves of spiral galaxies for three well known dark matter density profiles: Pseudo isothermal, Navarro-Frenk-White and Burkert dark matter density profiles. We estimate the brane tension parameter using a sample of high resolution observed rotation curves of low surface brightness spiral galaxies and a synthetic rotation curve for the three density profiles. Also, the fittings using the brane theory model of the rotation curves are compared with standard Newtonian models. We found that Navarro-Frenk-White model prefers lower values of the brane tension parameter, on the average $\lambda \sim 0.73 \times 10^{-3}$ eV^4, therefore showing clear brane effects. Burkert case does prefer higher values of the tension parameter, on the average $\lambda \sim 0.93$ eV^4 – 46 eV^4, i.e., negligible brane effects. Whereas pseudo isothermal is an intermediate case. Due to the low densities found in the galactic medium it is almost impossible to find evidence of the presence of extra dimensions. In this context, we found that our results show weaker bounds to the brane tension values in comparison with other bounds found previously, as the lower value found for dwarf stars composed of a polytropic equation of state, $\lambda \approx 10^4$ MeV^4.

1. Introduction

General Relativity (GR) is the cornerstone of astrophysics and cosmology, giving predictions with unprecedented success. At astrophysical scales GR has been tested in, for example, the solar system, stellar dynamics, black hole formation and evolution, among others (see for instance[1, 2, 3, 4, 5]). However, GR is being currently tested in the study of several phenomena that can be significant challenges to it, generating important changes never seen before. Dark matter (DM) and dark energy are undoubtedly ones of the major challenges of modern cosmology. They comprise approximately 27% for DM and 68% for dark energy, of our universe[6] allowing the formation of large scale structures[7, 8]. Dark matter has been invoked as the mechanism to stabilize spiral galaxies and to provide the matter distribution component that explain the observed rotation curves. Nowadays the best model of the universe we have is the *concordance* or ΛCDM model that has been successful in explaining the very large-scale structure formation, the statistics of the distribution of galaxy clusters, the temperature anisotropies of the cosmic microwave background radiation (CMB) and many other astronomical observations. In spite of all successes we have mentioned, this model has several problems, for example: predicts too much power on small scale[9], then it over predicts the number of observed satellite galaxies[10, 11] and predicts halo profiles that are denser and cuspier than those inferred observationally[12, 13], and also predicts a population of massive and concentrated subclass that are inconsistent with observations of the kinematics of Milky Way satellites[14].

One of the first astronomical observations that brought attention on DM was the observation of rotation curves of spiral galaxies by Rubin and coworkers[15], these observations turned out to be the main tool to investigate the role of DM at galactic scales: its role in determining the structure, how the mass is distributed, and the dynamics, evolution, and formation of spiral galaxies. Remarkably, the corresponding rotation curves of galaxies, can be explained with the density profiles of different Newtonian DM models like Pseudo Isothermal profile (PISO)[16], Navarro-Frenk-White profile (NFW)[12] or Burkert profile[17], among others[18]; except by the fact that until now it is unsolved the problem of cusp and core in the densities profiles. In this way, none of them have the last word because the main questions, about the density distribution and of course the *nature* of DM, has not been answered.

Alternative theories of gravity have been used to model DM. For instance a scalar field has been proposed to model DM[19, 20], and has been used to study rotation curves of spiral galaxies[21]. This scalar field is coupled minimally to the metric, however, scalar fields coupled non minimally to the metric have also been used to study DM[9, 22, 23, 24]. Equivalently $f(R)$ models exists in the literature that analyzes rotation curves[25].

On the other hand, one of the best candidates to extend GR is the brane theory, whose main characteristic is to add another dimension having a five dimensional bulk where it is embedded a four dimensional manifold called the brane[26, 27]. This model is characterized by the fact that the standard model of particles is confined in the brane and only the gravitational interaction can travel in the bulk[26, 27]. The assumption that the five dimensional Einstein's equations are valid, generates corrections in the four dimensional Einstein's equations confined in the brane bringing information from the extra dimension[28]. These extra corrections in the Einstein's equations can help us to elucidate and solve the problems that afflicts the modern cosmology and astrophysics[29, 30, 31, 32, 33, 34, 35, 36, 37, 38, 39, 40, 41, 42].

Before we start, let us mention here some experimental constraints on braneworld models, most of them about the so-called brane tension, λ, which appears explicitly as a free parameter in the corrections of the gravitational equations mentioned above. As a first example we have the measurements on the deviations from Newton's law of the gravitational interaction at small distances. It is reported that no deviation is observed for distances $l \gtrsim 0.1\,\mathrm{mm}$, which then implies a lower limit on the brane tension in the Randall-Sundrum II model (RSII): $\lambda > 1\,\mathrm{TeV}^4$[43, 44]; it is important to mention that these limits do not apply to the two-branes case of the Randall-Sundrum I model (RSI) (see [45] for details). Astrophysical studies, related to gravitational waves and stellar stability, constrain the brane tension to be $\lambda > 5 \times 10^8\,\mathrm{MeV}^4$[33, 46], whereas the existence of black hole X-ray binaries suggests that $l \lesssim 10^{-2}\mathrm{mm}$[45, 47, 48]. Finally, from cosmological observations, the requirement of successful nucleosynthesis provides the lower limit $\lambda > 1\,\mathrm{MeV}^4$, which is a much weaker limit as compared to other experiments (another cosmological tests can be seen in: Ref. [49, 50, 51]).

In fact, this paper is devoted to study the main observable of brane theory which is the brane tension, whose existence delimits between the four dimensional GR and its high energy corrections. We are given to the task of perform a Newtonian approximation of the modified Tolman-Oppenheimer-Volkoff (TOV) equation, maintaining the effective terms provided by branes which cause subtle differences in the traditional dynamics. In this way we test the theory at galactic scale using high resolution measurements of rotation curves of a sample of low surface brightness (LSB) galaxies without photometry[52] and a synthetic rotation curve built from 40 rotation curves of spirals of magnitude around $M_I = -18.5$ where was found that the baryonic components has a very small contribution[53], assuming PISO, NFW and Burkert DM profiles respectively and with that, we constraint the preferred value of brane tension with observables. That the sample has no photometry means that the galaxies are DM dominated and then we have only two parameters related to the distribution of DM, a density scale and a length scale, adding the brane tension we have three parameters in total to fit. The brane tension fitted values are compared among the traditional DM density profile models of spiral galaxies (PISO, NFW and Burkert) and against the same models without the presence of branes and confronted with other values of the tension parameter coming from cosmological and astrophysical observational data.

This paper is organized as follows: In Sec. 2 we show the equations of motion (modified TOV equations) for a spherical symmetry and the appropriate initial conditions. In Sec. 3 we explore the Newtonian limit and we show the mathematical expression of rotation velocity with brane modifications; particularly we show the modifications to rotation curve expressions of PISO, NFW and Burkert DM profiles and they are compared with models without branes. In Sec. 4 we test the DM models plus brane with observations: we use a sample of high resolution measurements of rotation curves of LSB galaxies and a synthetic rotation curve representative of 40 rotation curves of spirals where the baryonic component has a very small contribution. Finally in Sec. 5, we discuss the results obtained in the paper and make some conclusions.

In what follows, we work in units in which $c = \hbar = 1$, unless explicitly written.

2. Review of equations of motion for branes

Let us start by writing the equations of motion for galactic stability in a brane embedded in a five-dimensional bulk according to the RSII model[27]. Following an appropriate computation (for details see[45, 28]), it is possible to demonstrate that the modified four-dimensional Einstein's equations can be written as

$$G_{\mu\nu} + \xi_{\mu\nu} + \Lambda_{(4)} g_{\mu\nu} = \kappa_{(4)}^2 T_{\mu\nu} + \kappa_{(5)}^4 \Pi_{\mu\nu} + \kappa_{(5)}^2 F_{\mu\nu}, \tag{1}$$

where $\kappa_{(4)}$ and $\kappa_{(5)}$ are respectively the four and five- dimensional coupling constants, which are related as: $\kappa_{(4)}^2 = 8\pi G_N = \kappa_{(5)}^4 \lambda/6$, where λ is defined as the brane tension, and G_N is the Newton constant. For purposes of simplicity, we will not consider bulk matter, which translates into $F_{\mu\nu} = 0$, and discard the presence of the four-dimensional cosmological constant, $\Lambda_{(4)} = 0$, as we do not expect it to have any important effect at galactic scales (for a recent discussion about it see[54]). Additionally, we will neglect any nonlocal energy flux, which is not allowed by the static spherically symmetric solutions we will study below[33].

The energy-momentum tensor, the quadratic energy-momentum tensor, and the Weyl (traceless) contribution, have the explicit forms

$$T_{\mu\nu} = \rho u_\mu u_\nu + p h_{\mu\nu} \,, \tag{2a}$$

$$\Pi_{\mu\nu} = \frac{1}{12}\rho \left[\rho u_\mu u_\nu + (\rho + 2p) h_{\mu\nu}\right] \,, \tag{2b}$$

$$\xi_{\mu\nu} = -\frac{\kappa_{(5)}^4}{\kappa_{(4)}^4} \left[\mathcal{U} u_\mu u_\nu + \mathcal{P} r_\mu r_\nu + \frac{h_{\mu\nu}}{3}(\mathcal{U} - \mathcal{P})\right] . \tag{2c}$$

Here, p and ρ are, respectively, the pressure and energy density of the stellar matter of interest, $\mathcal{U}$ is the nonlocal energy density, and $\mathcal{P}$ is the nonlocal anisotropic stress. Also, u_α is the

IOP Conf. Series: Journal of Physics: Conf. Series **1010** (2018) 012006 doi:10.1088/1742-6596/1010/1/012006

four-velocity (that also satisfies the condition $g_{\mu\nu}u^\mu u^\nu = -1$), r_μ is a unit radial vector, and $h_{\mu\nu} = g_{\mu\nu} + u_\mu u_\nu$ is the projection operator orthogonal to u_μ.

Spherical symmetry indicates that the metric can be written as:

$$ds^2 = -B(r)dt^2 + A(r)dr^2 + r^2(d\theta^2 + \sin^2\theta d\varphi^2)\,. \tag{3}$$

We define the reduced Weyl functions $\mathcal{V} = 6\mathcal{U}/\kappa^4_{(4)}$, and $\mathcal{N} = 4\mathcal{P}/\kappa^4_{(4)}$. Also we define the effective mass as:

$$\mathcal{M}'_{eff} = 4\pi r^2 \rho_{eff}. \tag{4}$$

Then, from Eqs. (1) and (2) and after straightforward calculations we have the following equations of motion:

$$p' = -\frac{G_N}{r^2}\frac{4\pi\, p_{eff}\, r^3 + \mathcal{M}_{eff}}{1 - 2G_N\mathcal{M}_{eff}/r}(p+\rho)\,, \tag{5a}$$

$$\mathcal{V}' + 3\mathcal{N}' = -\frac{2G_N}{r^2}\frac{4\pi\, p_{eff}\, r^3 + \mathcal{M}_{eff}}{1 - 2G_N\mathcal{M}_{eff}/r}(2\mathcal{V} + 3\mathcal{N}) - \frac{9}{r}\mathcal{N} - 3(\rho + p)\rho'\,, \tag{5b}$$

where a prime indicates derivative with respect to r and define $A(r) = [1 - 2G_N\mathcal{M}(r)_{eff}/r]^{-1}$, and the effective energy density and pressure, respectively, are given as:

$$\rho_{eff} = \rho\left(1 + \frac{\rho}{2\lambda}\right) + \frac{\mathcal{V}}{\lambda}\,, \tag{6a}$$

$$p_{eff} = p\left(1 + \frac{\rho}{\lambda}\right) + \frac{\rho^2}{2\lambda} + \frac{\mathcal{V}}{3\lambda} + \frac{\mathcal{N}}{\lambda}\,. \tag{6b}$$

Even though we will not consider exterior galaxy solutions, we must anyway take into account the information provided by the Israel-Darmois (ID) matching condition, which for the case under study can be written as[33]:

$$(3/2)\rho^2(R) + \mathcal{V}^-(R) + 3\mathcal{N}^-(R) = 0\,. \tag{7}$$

In this case, the superscript $(-)$ indicates the interior value of the quantity at the halo surface[1] of the galaxy, assuming that $\rho(r > R) = 0$ where R denotes the maximum size of the galaxy. Also, the previous equation considers that the exterior must be given by the Schwarzschild metric and in general the following condition must be fulfilled $\mathcal{V}(r \geq R) = 0 = \mathcal{N}(r \geq R)$ (see[36] for details).

For completeness, we just note that the exterior solutions of the metric functions are given by the well known expressions $B(r) = A^{-1}(r) = 1 - 2G_N M_{eff}/r$.

Finally, we impose $\mathcal{N} = 0$ (see[36]). Implying that Eq. (7) is restricted as:

$$-(3/2)\rho^2(R) = \mathcal{V}^-(R)\,, \tag{8}$$

with the aim of maintain a galaxy Schwarzschild exterior.

[1] We denote the surface of the galaxy as a region where does not exist DM or baryons, *i.e.*, the intergalactic space.

IOP Conf. Series: Journal of Physics: Conf. Series **1010** (2018) 012006 doi:10.1088/1742-6596/1010/1/012006

3. Low energy limit and rotation curves

To begin with, we observe, from Eq. (5a) in the low energy (Newtonian) limit, that we have: $r^2 p' = -G_N \mathcal{M}_{eff}\rho$. Differentiating we found

$$\frac{d}{dr}\left(\frac{r^2}{\rho}\frac{dp}{dr}\right) = -4\pi r^2 G_N \rho_{\rm eff}. \tag{9}$$

From here it is possible to note that $d\Phi/dr = -\rho^{-1}(dp/dr)$ resulting in

$$\nabla^2\Phi_{\rm eff} = \frac{1}{r^2}\frac{d}{dr}\left(r^2\frac{d\Phi_{\rm eff}}{dr}\right) = 4\pi G_N \rho_{\rm eff}, \tag{10}$$

where it is necessary to define the energy density of the DM together with the nonlocal energy density. Notice that the nonlocal energy density can be obtained easily from Eq. (5b) in the galaxy interior and also the fluid behaves like dust, implying the condition $p = 0$, always fulfilling the low energy condition $4\pi r^3 p_{eff} \ll \mathcal{M}_{eff}$ and $2G_N\mathcal{M}_{eff}/r \ll 1$, between effective quantities and in consequence $4G_N\mathcal{M}_{eff}\mathcal{V}/r^2 \sim 0$, is negligible.

In addition, the rotation curve is obtained from the contribution of the effective potential. This expression can be written as:

$$\begin{aligned} V^2(r) &= r\left|\frac{d\Phi_{\rm eff}}{dr}\right| = \frac{G_N\mathcal{M}_{eff}(r)}{r} \\ &= \frac{G_N}{r}\left[\mathcal{M}_{DM}(r) + \mathcal{M}_{Brane}(r)\right], \end{aligned} \tag{11}$$

where $\mathcal{M}_{DM}(r)$ is the contribution to the mass from DM, $\mathcal{M}_{Brane}(r)$ gives the modification to the DM mass that comes from the brane; and $\mathcal{M}_{eff}(r)$ must be greater than zero. From here, it is possible to study the rotation curve of the DM, assuming a variety of density profiles.

Before we start let us define the following dimensionless variables: $\bar{r} \equiv r/r_{\rm s}$, $v_0^2 \equiv 4\pi G_N r_{\rm s}^2 \rho_{\rm s}$ and $\bar{\rho} \equiv \rho_{\rm s}/2\lambda$ where $\rho_{\rm s}$, is the central density of the halo and r_s is associated with the central radius of the halo.

3.1. Pseudo isothermal profile for dark matter

Here we consider that DM density profile is given by the PISO[16] model written as:

$$\rho_{\rm PISO}(r) = \frac{\rho_{\rm s}}{1+\bar{r}^2}. \tag{12}$$

From Eq. (11), together with Eq. (12), it is possible to obtain:

$$\begin{aligned} V^2_{\rm PISO}(\bar{r}) &= v_0^2\left\{\left(1 - \frac{1}{\bar{r}}\arctan\bar{r}\right)\right. \\ &\left. +\bar{\rho}\left(\frac{1}{1+\bar{r}^2} - \frac{1}{\bar{r}}\arctan\bar{r}\right)\right\}. \end{aligned} \tag{13}$$

In the limit $\bar{\rho} \to 0$, we recover the classical rotation curve associated with PISO for DM. The effective density must be positive defined, then $\lambda > \rho_s$ must be fulfilled. The first right-hand term in parenthesis in Eq. (13) is PISO dark matter contribution and the second is brane's contribution.

3.2. Navarro-Frenk-White profile for dark matter

Another interesting case (motivated by cosmological N-body simulations) is the NFW density profile, which is given by[55]:

$$\rho_{\rm NFW}(r) = \frac{\rho_{\rm s}}{\bar{r}(1+\bar{r})^2}. \tag{14}$$

This is a density profile that diverges as $r \to 0$ and it is not possible to say that ρ_s is related with the central density of the DM distribution. Also density goes as $1/\bar{r}^3$ when $\bar{r} \gg 1$. However, in this particular case, we will still be calling them the *central* density and radius of the NFW matter distribution. From Eq. (11), together with Eq. (14) we obtain the following rotation curve:

$$\begin{aligned} V^2_{\rm NFW}(\bar{r}) &= v_0^2 \left\{ \left(\frac{(1+\bar{r})\ln(1+\bar{r}) - \bar{r}}{\bar{r}(1+\bar{r})} \right) \right. \\ &\left. + \frac{2\bar{\rho}}{3\bar{r}} \left(\frac{1}{(1+\bar{r})^3} - 1 \right) \right\}. \end{aligned} \tag{15}$$

The first right-hand term in parenthesis in Eq. (15) is NFW dark matter contribution and the second one is the brane's contribution. Notice that we recover also the classical limit when $\bar{\rho} \to 0$.

In addition, it is important to remark that the effective density must be positive defined, then $\lambda > \rho_s r_s / r$. Also, if $\mathcal{M}(r)$ must be greater than zero, then $r > r_{min}$ where r_{min} is given by the solution to the following equation:

$$\frac{2}{3}\bar{\rho} = \frac{(\alpha+1)^2[(\alpha+1)\ln(\alpha+1) - \alpha]}{(\alpha+1)^3 - 1}, \tag{16}$$

where we have defined $\alpha \equiv r_{min}/r_s$ as a dimensionless quantity.

3.3. Burkert density profile for dark matter

Another density profile was proposed by Burkert[17], which it has the form:

$$\rho_{\rm Burk} = \frac{\rho_{\rm s}}{(1+\bar{r})(1+\bar{r}^2)}. \tag{17}$$

Again, from Eq. (11), together with Eq. (17) we obtain the following rotation curve:

$$\begin{aligned} V^2_{\rm Burk}(\bar{r}) &= \frac{v_0^2}{4\bar{r}} \left\{ \left(\ln[(1+\bar{r}^2)(1+\bar{r})^2] - 2\arctan(\bar{r}) \right) \right. \\ &\left. + \frac{1}{2}\bar{\rho} \left(\frac{1}{1+\bar{r}} + \frac{1}{1+\bar{r}^2} + \arctan(\bar{r}) - 2 \right) \right\}. \end{aligned} \tag{18}$$

In the limit $\bar{\rho} \to 0$, we recover the classical rotation velocity associated with Burkert density profile[17]. The effective density must be positive defined, then $\lambda > \rho_s$. Again the first right-hand term in parenthesis in Eq. (18) is Burkert DM contribution and the second one comes from the brane's contribution.

3.4. A general model for a DM density profile for galaxies

A generic model for DM in galaxies has been proposed by Zhao[56] and extended by Spano *et al.*[57]:

$$\rho_Z(r) = \frac{\rho_s}{(c + (r/r_s)^{-\gamma})(1 + (r/r_s)^{\alpha})^{(\beta+\gamma)/\alpha}} \tag{19}$$

where the set of parameters: $(c, \alpha, \beta, \gamma)$ characterises a particular model. We can, with this model, produce the above models. For instance, $(0,2,2,0)$ gives us PISO; $(0,1,3,-1)$ is for NFW and $(1,2,3,-1)$ gives Burkert model. Therefore a combined data analysis can be done with this more general model. This analysis is in progress and will be reported elsewhere.

Galaxy sample						
Observational data						
Galaxy	Morphology	D (Mpc)	$V_{\rm hel}$ (km/s)	$M_{\rm abs}(B)$ (mag)	$R_{\rm max}$ (kpc)	$V_{\rm max}$ (km/s)
(1)	(2)	(3)	(4)	(5)	(6)	(7)
ESO 0140040	Spiral	212	16064	-21.6	29.2	263
ESO 2060140	Spiral	60	4704	-19.2	11.6	118
ESO 3020120	Spiral, hint of bar	69	5311	-19.1	11.0	86
ESO 3050090	Barred spiral	11	1019	-17.3	4.8	54
ESO 4250180	Spiral	86	6637	-20.5	14.4	145
ESO 4880049	Inclined Magellanic bar	22	1800	-16.8	6.0	97
F570-V1	Spiral	144	10714	-19.2	11.9	145
U 11454	Fuzzy spiral, small core	91	6628	-18.6	11.9	152
U 11616	Fuzzy, irregular	73	5244	-20.3	9.6	143
U 11648	Irregular	48	3350	-21.0	12.7	145
U 11748	Irregular, bright core/bar?	73	5265	-22.9	21.0	242
U 11819	Fuzzy	60	4261	-20.3	11.7	153

Table 1. Sample of LSB galaxies without photometry as given in Ref. [52]. In column (1) we give the name of each galaxy; in column (2) we give the morphology of each galaxy; column (3) is the distance to the galaxy in Mpc; column (4) is the heliocentric systemic velocity; column (5) is the absolute magnitude in B-band of each galaxy; column (6) is distance in kpc where the maximum velocity is reached; column (7) is the maximum velocity of each rotation curve.

4. Brane's tension from fitting rotation curves

To start with the data analysis, we best fit the sample of observational rotation curves by minimizing the χ^2:

$$\chi^2 = \sum_{i=1}^{N} \left(\frac{V_{theo} - V_{exp\ i}}{\delta V_{exp\ i}} \right)^2 , \tag{20}$$

where i runs from one up to the number of points in the data, N; V_{theo}, is computed according to the velocity profile under consideration and $\delta V_{exp\ i}$, is the error in the measurement of the rotation velocity. Notice that the free parameters are for DM-Branes: r_s, ρ_s and λ. In the tables below we show $\chi^2_{red} \equiv \chi^2/(N - n_p - 1)$ where n_p is the number of parameters to fit, in our case, $n_p = 3$.

The analysed sample of galaxies (shown in Table 1) are twelve high resolution rotation curves of LSB galaxies without photometry (visible components, such as gas and stars, are negligible) as given in Ref. [52]. It is a subsample of the sample II in this reference and quoted as galaxies without photometry. The sample of analysed galaxies is shown in Table 1. In column (6) it is shown the radius at which the maximum velocity is reached for each galaxy, at this radius one expects DM is the dominant component. We select this subsample of the sample II in Ref. [52] in such a way to have a wide range of velocity values, as can be seen in column (7) that is correlated with the magnitude of the galaxies, column (5). In this manner we have an appropriate sample with different types of galaxies to test the brane's tension. We remark that in this part we use units such that $4\pi G_N = 1$, velocities are in km/s, and distances are given in kpc.

4.1. Results: PISO profile + Branes

We have estimated the parameters of the PISO+branes model and compared with PISO model without brane contribution, minimizing the appropriate χ^2 for the sample of observed rotation curves, using Eq. (20) with Eq. (13) and taking into account that $\lambda > \rho_s$ must be fulfilled.

In Fig. 1 we show, for each one of the galaxies in the sample, the plots of the PISO theoretical rotation curve (solid line), that best fit of the corresponding observational data (orange symbols); also shown are the errors of the estimation (brown symbols). For each galaxy we have plotted the contribution to the rotation velocity due only to the brane (red long-dashed curve) and only to the dark matter PISO density profile (blue short-dashed curve), see Eq. (13). Brane effects are very clear in galaxies: ESO 2060140, ESO 3020120, U 11616, U 11648, U 11748, U 11819. In Table 2 it is shown the central density, central radius and the brane tension which is the free parameter of the brane theory (only in PISO+branes). As a comparison, it is also shown the central density and radius without brane contribution. The worst fitted galaxies (high χ^2_{red} value) were: U 11648, U 11748. The fitted brane tension values presents great dispersion, from the lower value: 0.167 $M_\odot/\text{pc}^3$, ESO 3020120 to the higher value: 108.096 $M_\odot/\text{pc}^3$, ESO 4880049. It is useful to have λ in eV, where the conversion from solar masses to eV is: $1 M_\odot/\text{pc}^3 \sim 2.915 \times 10^{-4}\text{eV}^4$. The brane tension parameter has an average value of $\langle\lambda\rangle_{\text{PISO}} = 33.178\ M_\odot/\text{pc}^3$ with a standard deviation $\sigma_{\text{PISO}} = 40.935\ M_\odot/\text{pc}^3$. Notice that we can not see a clear tendency to a λ value or range of values.

4.2. Results: NFW profile + Branes

For the NFW density profile case we have the following results: We have estimated parameters with and without brane contribution by minimizing the corresponding χ^2, Eq. (20) with Eq. (15), for the sample of observed rotation curves and taking into account that $\lambda > \rho_s r_s/r$ in order to have an effective density positive defined, always fulfilling Eq. (16).

In Fig. 2, it is shown, for each galaxy in the sample of the LSB galaxies, the theoretical fitted curve to a preferred brane tension value (solid line), the NFW curve and the rotation curve associated with the mass lost by the effects of branes, see Eq. (15). In Table 3 it is shown, for the sample, the central density, central radius and χ^2_{red} values without branes; and the central density, central radius, brane tension and χ^2_{red} values with branes contribution. Galaxy U 11748 is the worst fitted case with $\chi^2_{red} = 2.163$. For galaxies: ESO 4250180, ESO 4880049, and U 11648, there are not clear brane effects. Galaxy U 11648 is an *outlier* with a brane tension value of 4323.28 $M_\odot/\text{pc}^3$ that is out of the range of preferred values of the other galaxies in the sample. Notice that we have found a preferred range of tension values, from 0.487 to 9.232 $M_\odot/\text{pc}^3$. Without the outlier, the brane tension parameter has an average value of $\langle\lambda\rangle_{\text{NFW}} \simeq 2.51\ M_\odot/\text{pc}^3$ with a standard deviation $\sigma_{\text{NFW}} \simeq 3.015\ M_\odot/\text{pc}^3$.

4.3. Results: Burkert+Branes profile

In the case of Burkert DM density profile, we have also estimated the parameters of the Burkert+branes model and they were compared with Burkert model without branes, minimizing the appropriate χ^2_{red}, Eq. (20) with Eq. (18), for the sample of observed rotation curves. We have considered that $\lambda > \rho_s$ must be fulfilled.

The results are shown in Fig. 3, where it is plotted the fit to a preferred brane tension value, remarking the total rotation curve (solid line), the Burkert DM density contribution curve (blue short-dashed line) and the rotation curve associated with the mass lost by the effects of branes (red dashed line), see Eq. (18). In Table 4 it is shown the fitted values for the central density,

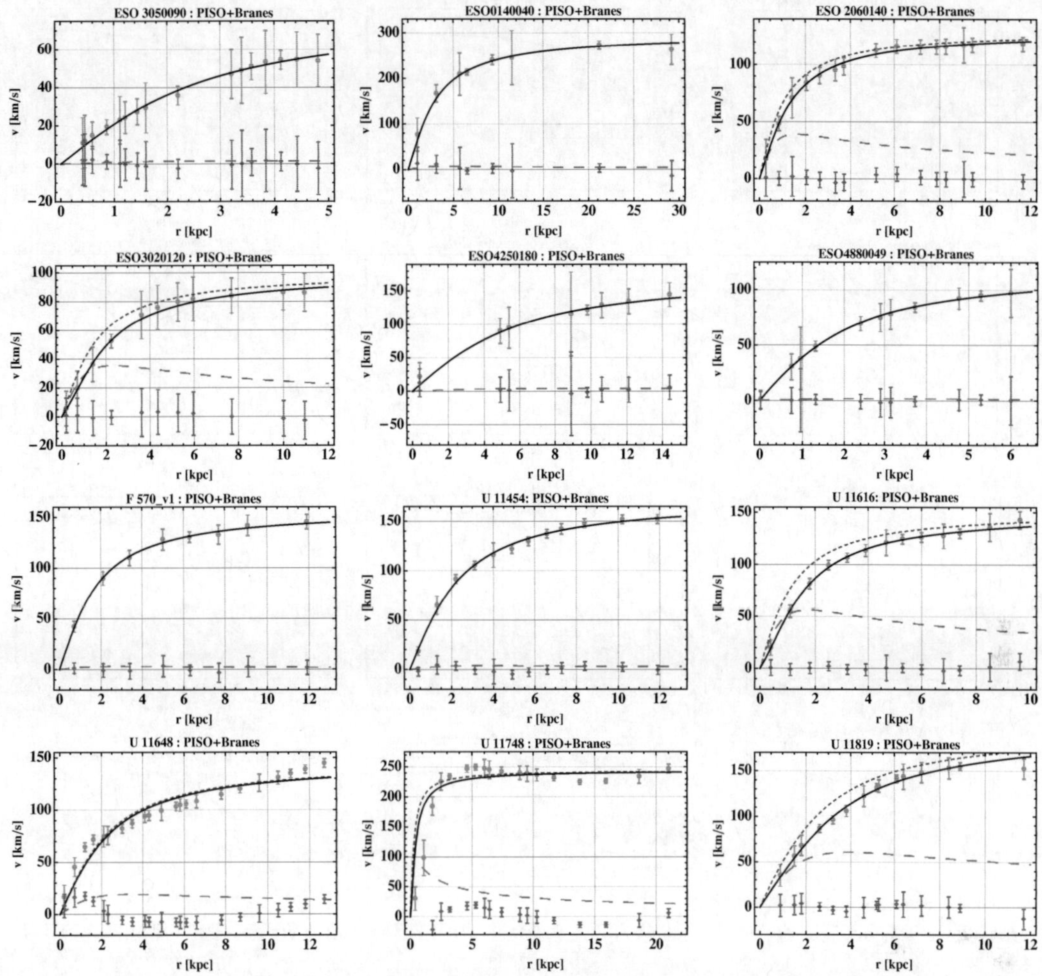

Figure 1. Group of analyzed galaxies using the modified rotation curve for PISO profile: ESO 3050090, ESO 0140040, ESO 2060140, ESO 3020120, ESO 4250180, ESO 4880049, 570_V1, U11454, U11616, U11648, U11748, U11819. We show in the plots: Total rotation curve (solid black line), only PISO curve (short dashed blue curve) and the rotation curve associated with the mass lost by the effect of the brane (red dashed curve).

central radius and the corresponding value of the χ^2_{red} without brane contribution; and the fitted values for the central density, central radius, brane tension, and theirs χ^2_{red} values with brane contribution. The worst fitted (high values of χ^2_{red}) galaxies are: U 11648 and U 11748. Galaxies ESO 3020120, U 11748, and U 11819 show a clear brane effects and also are outliers. The main tendency is that λ has values of the order of $10^3\ M_\odot/\mathrm{pc}^3$ or above, approximately. The brane tension parameter, without the outliers, for the DM Burkert profile case has an average value of $\langle\lambda\rangle_{\mathrm{Burk}} \simeq 3192.02\ M_\odot/\mathrm{pc}^3$, and a standard deviation of $\sigma_{\mathrm{Burk}} \simeq 2174.97\ M_\odot/\mathrm{pc}^3$.

4.4. Results: a synthetic rotation curve

Finally, we show the fitting results of the DM models plus brane's contribution to a synthetic rotation curve. This synthetic rotation curve was made of 40 rotation curves of galaxies with magnitudes around $M_I = -18.5$[53]. These 40 rotation curves came out of 1100 galaxies that

IOP Conf. Series: Journal of Physics: Conf. Series **1010** (2018) 012006 doi:10.1088/1742-6596/1010/1/012006

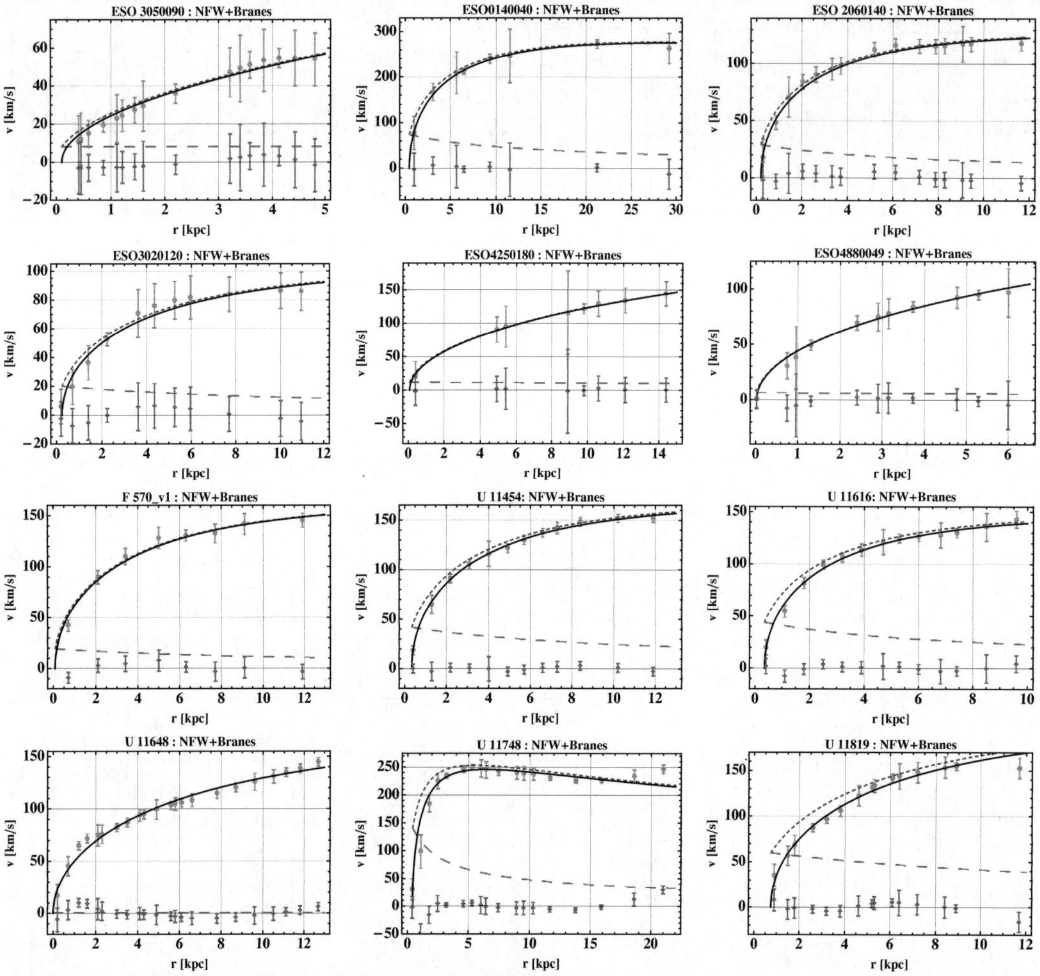

Figure 2. Group of analyzed galaxies using the modified rotation curve for NFW profile: ESO 3050090, ESO 0140040, ESO 2060140, ESO 3020120, ESO 4250180, ESO 4880049, 570_V1, U11454, U11616, U11648, U11748, U11819. We show in the plots: Total rotation curve (solid black line), only NFW curve (short dashed blue curve) and the rotation curve associate with the mass lost by the effect of the brane (red dashed curve).

gave the universal rotation curve for spirals. For this sample of low luminosity galaxies, of $M_I = -18.5$, it was shown that the baryonic disk has a very small contribution (for details see reference[53]).

In this subsection we are now using units such $G = R_{opt} = V(R_{opt}) = 1$, where R_{opt} and $V(R_{opt})$ are the optical radius and the velocity at the optical radius, respectively. R_{opt} is the radius encompassing 83 per cent of the total integrated light. For an exponential disk with a surface brightness given by: $I(r) \propto \exp(-r/R_D)$, we have that $R_{opt} = 3.2R_D$[53].

In figure 4 we show the synthetic rotation curve and the fitting results using PISO, NFW and Burkert profiles with and without brane's contribution.

As we can see in Table 5 the same trend is observed in the brane's tension values as compared with the results for the LSB catalog analyzed above using PISO, NFW and Burkert as a DM profiles: lower value is obtained for NFW model and higher value is obtained for Burkert density

IOP Conf. Series: Journal of Physics: Conf. Series **1010** (2018) 012006 doi:10.1088/1742-6596/1010/1/012006

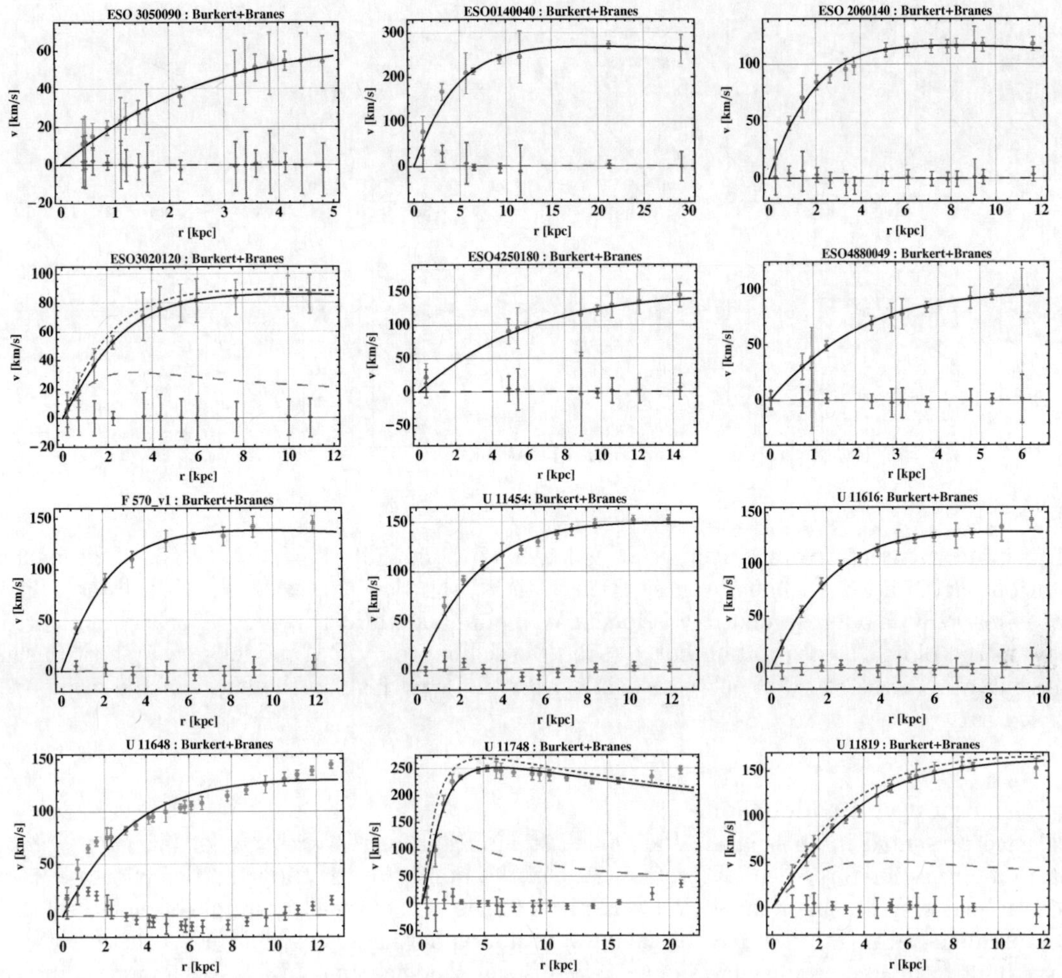

Figure 3. Group of analyzed galaxies using the modified rotation curve for Burkert profile: ESO 3050090, ESO 0140040, ESO 2060140, ESO 3020120, ESO 4250180, ESO 4880049, 570_V1, U11454, U11616, U11648, U11748, U11819. We show in the plots: Total rotation curve (solid black line), only Burkert curve (short dashed blue curve) and the rotation curve associate with the mass lost by the effect of the brane (red dashed curve).

profile.

Given that this synthetic rotation curve is built from 40 rotation curves of real spirals, the values of the brane's tension in table 5 is representative of all these rotation curves. Then, for PISO model $\lambda = 60.692\ M_{\odot}/\mathrm{pc}^3$, a value that is greater than the average value of the tension shown in Table 2 but inside the interval marked by the standard deviation. For NFW model $\lambda = 226.054\ M_{\odot}/\mathrm{pc}^3$, this value is lower than the average value reported in Table 3 and inside the range marked by the standard deviation. For Burkert model $\lambda = 1.58 \times 10^5\ M_{\odot}/\mathrm{pc}^3$ this value is well above than the average value shown in Table 4; a value outside the range marked by the standard deviation.

IOP Conf. Series: Journal of Physics: Conf. Series **1010** (2018) 012006 doi:10.1088/1742-6596/1010/1/012006

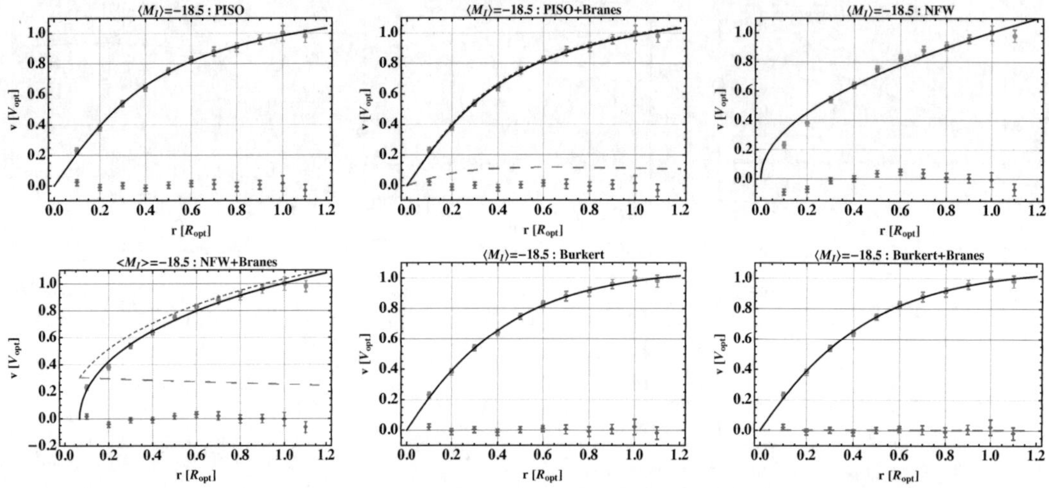

Figure 4. Synthetic rotation curve of galaxies with magnitud $M_I = -18.5$. Left panels: rotation curves fitted without branes. Right panels: rotation curves fitted with branes. First row is for PISO model; second row is for NFW model and third row is for Burkert model. We show in the plots: Total rotation curve (solid black line), only DM model curve (short dashed blue curve) and the rotation curve associate with the mass lost by the effect of the brane (red dashed curve).

5. Discussion and conclusions

We have presented in this paper, the effects steaming from the presence of branes in galaxy rotation curves for three density profiles used in the literature to study the behavior of DM at galactic scales. With this in mind, we were given to the task of study a sample of high resolution measurements of rotation curves of galaxies without photometry[52] and a synthetic rotation curve built from 40 rotation curves of galaxies of magnitude around $M_I = -18.5$, and fitting the values of ρ_s, r_s and λ through minimizing the χ^2 value and we have compared with the standard results of ρ_s, r_s for each DM density profile without branes. The results for every observable in the three different profiles were summarized and compared in Tables 2-5.

From here, it is possible to observe how the results show a weaker limit for the value of brane tension ($\sim 10^{-3}$ eV^4 – 46 eV^4) for the three models, in comparison with other astrophysical and cosmological studies[43, 44, 45, 33, 46, 47, 48, 49, 50, 51]; for example, Linares *et al.* [41] show that weaker values than $\lambda \simeq 10^4$ MeV^4, present anomalous behavior in the compactness of a dwarf star composed by a polytropic EoS, concluding that a wide region of their bound will show a non compactness stellar configuration, if it is applied to the study shown in [41].

It is important to notice that chosen a value of brane tension that not fulfill our bounds imposed through the paper, generate an anomalous behavior in the center of the galaxy which is characteristic of the model. Remarkable, for higher values of this bound, the modified rotation curves are in good agreement with the observed rotation curves of the samples that we use, presenting only the distinctive features of each density profile: For example, NFW dark matter density profile prefers lower values of the brane tension (on the average $\lambda \sim 0.73 \times 10^{-3}$ eV^4), implying clear effects of the brane; PISO dark matter case has an average value of $\lambda \sim 0.96 \times 10^{-2}$ eV^4 and show relatively the maximum dispersion on the fitted values of the brane tension; whereas Burkert DM density profile shows negligible brane effects, on the average $\lambda \sim 0.93$ eV^4 – 46 eV^4. Therefore, if the tension of the brane is supposed to play a noticeable role on galactic dynamic, it has to have values of the order of meV^4 but the corresponding density profile will be

	PISO			PISO + branes			
Galaxy	ρ_s ($M_\odot$/pc^3)	r_s (kpc)	χ^2_{red}	ρ_s ($M_\odot$/pc^3)	r_s (kpc)	λ ($M_\odot$/pc^3)	χ^2_{red}
ESO 3050090	0.02732	2.0911	0.052	0.02738	2.0889	13.148	0.056
ESO 0140040	0.24930	2.5588	0.216	0.25012	2.5546	77.901	0.271
ESO 2060140	0.23308	1.1638	0.115	0.39715	0.8863	0.787	0.121
ESO 3020120	0.05420	1.8953	0.038	0.11611	1.2721	0.167	0.031
ESO 4250180	0.03019	4.3977	0.114	0.03028	4.3915	10.597	0.143
ESO 4880049	0.10235	1.6224	0.019	0.10245	1.6217	108.096	0.022
F 570_V1	0.21403	1.4723	0.100	0.21446	1.4708	107.042	0.124
U 11454	0.15115	1.9263	0.426	0.15155	1.9238	58.574	0.478
U 11616	0.19931	1.4978	0.188	0.45719	0.9685	0.600	0.141
U 11648	0.10588	1.9476	3.936	0.11608	1.8602	1.244	4.167
U 11748	8.16533	0.3674	6.101	15.6937	0.2652	19.680	6.303
U 11819	0.08888	2.9326	0.342	0.15733	2.1650	0.302	0.355

Table 2. Parameters of the analyzed sample with PISO rotation velocity. From left to right, the columns read: name of the galaxy; central density in units of $M_\odot/\text{pc}^3$; central radius in kpc and the χ^2_{red} value. And parameters of the analyzed sample with PISO+Branes rotation velocity: central density in units of $M_\odot/\text{pc}^3$; central radius in kpc; brane tension in units of $M_\odot/\text{pc}^3$ and the χ^2_{red} value. It is useful to have λ in eV, the conversion from solar masses to eV is: $M_\odot/\text{pc}^3 \sim 2.915 \times 10^{-4}\text{eV}^4$. Note that the average value of the fitted values of brane tension is $\langle\lambda\rangle_{\text{PISO}} \simeq 33.178\ M_\odot/\text{pc}^3$ with a standard deviation $\sigma_{\text{PISO}} \simeq 40.935\ M_\odot/\text{pc}^3$.

NFW, a cuspy halo. On the contrary for core density profile brane's effect are not significant.

Even more, let us go further and discuss briefly the changes caused by the presence of branes in the problem of cusp/core. Notice that the part that plays a role in the rotation curve is the effective density which is written in terms of brane corrections as: $\rho_{eff} = \rho(1 - \rho/\lambda)$; it is notorious how the small perturbations alleviate the cusp problem which afflicts NFW, albeit the excessive presence of these terms could generates a negative effective density profile; also PISO and Burkert show modifications when $r \to 0$ but does not pledge its core behavior while the brane tension only takes small values. In this way, the possibility of having a core behavior, help us to constraint the value of brane tension and still keep in the game the NFW profile.

Summarizing, it is really challenging to establish bounds in a dynamical systems like rotation curves in galaxies due to the low densities found in the galactic medium, giving only a weaker limits in comparison with other studies in a most energetic systems. Our most important conclusion, is that despite the efforts, we think that it is not straightforward to do that the results fit with other astrophysical and cosmological studies, being impractical and not feasible to find evidence of extra dimensions in galactic dynamics through the determination of the brane tension value, even more, we think that exist too much dispersion in the fitted values of the brane tension using this method for some DM density profile models. Also it is important to note that the value of the brane tension is strongly dependent on the characteristics of the galaxy studied, suggesting an average for the preferred value of the brane tension in each case. In addition, we notice that the effects of extra dimensions are stronger in the galactic core, suggesting that the NFW model is not appropriate in the search of constraints in brane theory due to the divergence in the center of the galaxy (see Eq. (16)); PISO and Burkert could be good candidates to explore the galactic core in this framework; however it is necessary a more extensive study before we obtain a definitive conclusion.

As a final note, we know that it is necessary to recollect more observational data to constraint the models or even give the final conclusion about extra dimensions (for or against), supporting the brane constraints shown through this paper with a more profound study of galactic dynamic or other tests like cosmological evidences presented in CMB anisotropies. However, this work is in progress and will be reported elsewhere.

	NFW			NFW + branes			
Galaxy	ρ_s ($M_\odot$/pc^3)	r_s (kpc)	χ^2_{red}	ρ_s ($M_\odot$/pc^3)	r_s (kpc)	λ ($M_\odot$/pc^3)	χ^2_{red}
ESO 3050090	0.00003	907.272	0.238	0.00002	1385.81	0.489	0.164
ESO 0140040	0.02548	16.148	0.170	0.03272	14.182	1.995	0.104
ESO 2060140	0.01989	8.110	0.455	0.01997	8.214	1.640	0.339
ESO 3020120	0.00262	19.720	0.369	0.00455	13.763	0.513	0.191
ESO 4250180	0.00052	119.413	0.015	0.00069	94.037	1.703	0.011
ESO 4880049	0.00134	54.811	0.185	0.00165	45.696	7.662	0.141
F 570_V1	0.01064	14.707	1.232	0.01325	12.799	4.122	1.140
U 11454	0.00797	19.013	3.208	0.01731	11.607	1.175	0.227
U 11616	0.01167	13.811	1.606	0.02679	8.221	1.245	0.384
U 11648	0.00404	24.409	0.990	0.00404	24.408	4323.28	1.045
U 11748	0.49834	3.191	3.782	0.82773	2.574	9.232	2.163
U 11819	0.00240	52.691	1.514	0.01033	18.102	0.487	0.690

Table 3. Parameters of the analyzed sample with NFW rotation velocity. From left to right, the columns read: name of the galaxy; central density in units of $M_\odot/\text{pc}^3$; central radius in kpc and the χ^2_{red} value. And properties and parameters of the analyzed sample with NFW+Branes rotation velocity: central density in units of $M_\odot/\text{pc}^3$; central radius in kpc; brane tension in units of $M_\odot/\text{pc}^3$ and the χ^2_{red} value. Note that the average value of the fitted values of brane tension is $\langle\lambda\rangle_{\text{NFW}} \simeq 362.795\ M_\odot/\text{pc}^3$ with a standard deviation $\sigma_{\text{NFW}} \simeq 1194.134\ M_\odot/\text{pc}^3$. See the text for more details about the average without the outliers galaxies.

	Burkert			Burkert + branes			
Galaxy	ρ_s ($M_\odot$/pc^3)	r_s (kpc)	χ^2_{red}	ρ_s ($M_\odot$/pc^3)	r_s (kpc)	λ ($M_\odot$/pc^3)	χ^2_{red}
ESO 3050090	0.03069	3.376	0.045	0.03069	3.376	3139.12	0.049
ESO 0140040	0.17489	5.973	0.667	0.17490	5.973	4545.85	0.834
ESO 2060140	0.19000	2.480	0.209	0.19002	2.480	5995.5	0.228
ESO 3020120	0.05668	3.372	0.007	0.08724	2.798	0.174	0.002
ESO 4250180	0.03071	7.695	0.134	0.03071	7.695	441.653	0.167
ESO 4880049	0.11095	2.729	0.051	0.11095	2.728	1682.83	0.059
F 570_V1	0.19128	2.391	0.683	0.19129	2.931	5268.5	0.854
U 11454	0.13899	3.705	1.090	0.13900	3.705	1203.16	1.227
U 11616	0.19596	2.751	0.359	0.19598	2.751	985.585	0.399
U 11648	0.08180	4.217	5.737	0.08180	4.217	5465.98	6.055
U 11748	1.43835	1.932	2.309	2.56134	1.555	2.777	2.429
U 11819	0.09828	4.843	0.204	0.13497	4.189	0.364	0.207

Table 4. Parameters of the analyzed sample with Burkert rotation velocity. From left to right, the columns read: name of the galaxy; central density in units of $M_\odot/\text{pc}^3$; central radius in kpc and the χ^2_{red} value. And parameters of the analyzed sample with Burkert+Branes rotation velocity: central density in units of $M_\odot/\text{pc}^3$; central radius in kpc; brane tension in units of $M_\odot/\text{pc}^3$ and the χ^2_{red} value. Note that the average value of the fitted values of brane tension is $\langle\lambda\rangle_{\text{Burk}} \simeq 2394.060\ M_\odot/\text{pc}^3$, $\sigma_{\text{Burk}} \simeq 2250.061\ M_\odot/\text{pc}^3$. See the text for more details about the average without the outliers galaxies.

DM profile	ρ_s ($M_\odot$/pc^3)	r_s (kpc)	χ^2_{red}	ρ_s ($M_\odot$/pc^3)	r_s (kpc)	λ ($M_\odot$/pc^3)	χ^2_{red}
PISO	2.637	0.968	0.447	2.761	0.945	60.692	0.511
NFW	0.0044	244.27	6.019	0.123	12.064	226.054	1.711
Burkert	2.958	1.571	0.321	2.958	1.571	1.58×10^5	0.367

Table 5. Parameters of the analyzed synthetic rotation curve with PISO, NFW and Burkert rotation velocity. From left to right, the columns read: name of the DM density profile; central density; central radius and the χ^2_{red} value. And parameters of the analyzed synthetic rotation curve with DM profile+Branes rotation velocity: central density; central radius; brane tension and the χ^2_{red} value.

IOP Conf. Series: Journal of Physics: Conf. Series **1010** (2018) 012006 doi:10.1088/1742-6596/1010/1/012006

Acknowledgments
MAG-A acknowledge support from SNI-México and CONACyT research fellow. Instituto Avanzado de Cosmología (IAC) collaboration.

References

[1] Fischbach E and Talmadge C 1999 *The search for Non-Newtonian Gravity* (Springer-Verlag, New York)
[2] Will C M 1993 *Theory and experiment in gravitational physics* (Cambridge University Press)
[3] Kamionkowski M 2007 (*Preprint* 0706.2986)
[4] Roos M 2010 (*Preprint* 1001.0316)
[5] Peebles P 2013 (*Preprint* 1305.6859)
[6] Planck Collaboration *et al.* *arXiv:1303.5076*
[7] Frenk C S and White S D M 2012 *Ann. Phys.* **507** (*Preprint* 1210.0544v1)
[8] Diaferio A 2008 *From the vacuum to the universe. Proceedings, Symposium, Innsbruck, Austria, October 19-20,2007* pp 71–85 (*Preprint* 0802.2532) URL http://inspirehep.net/record/779626/files/arXiv:0802.2532.pdf
[9] Rodríguez-Meza M A 2012 *Advances in Astronomy* **2012** 509682 (*Preprint* arXiv: 1112.5201)
[10] Moore B, Ghigna S, Governato F, Lake G, Quinn T, Stadel J and Tozzi P 1999 *ApJ* **524** L19
[11] Papastergis E, Martin A, Giovanelli R and Haynes M 2011 *ApJ* **739** 38
[12] Navarro J, Frenk C and White S 1997 *ApJ* **490** 493
[13] K Subramanian R C and Ostriker J 2000 *ApJ* **538** 528
[14] Boylan-Kolchin M, Bullock J and Kaplinghat M 2012 *MNRAS* **422** 1203
[15] Sofue Y and Rubin V 2001 *Annu. Rev. Astron. Astrophys.* **39** 137
[16] Begeman K G, Broeils A H and Sanders R H 1991 *Mon. Not. R. astr. Soc.* **294**
[17] Burkert A 1995 *Astrophys. J. Lett.* **L25**
[18] Einasto J 1965 *Trudy Inst. Astrofiz. Alma-Ata* **87**
[19] Dick R 1996 *Preprint hep-th/9609190*
[20] Cho Y M and Keum Y Y 1998 *Class. Quantum Grav.* **15** 907
[21] Guzmán F S and Matos T 2000 *Class. Quantum Grav.* **17** L9
[22] Rodriguez-Meza M A, Klapp J, Cervantes-Cota J L and Dehnen H 2001 *The influence of scalar fields in protogalactic interactions* (Kluwer Academic Publishers) p 213 (*Preprint* arXiv:0907.2466)
[23] Rodriguez-Meza M A and Cervantes-Cota J L 2004 *Mon. Not. Roy. Astron. Soc.* **350** 671
[24] Rodríguez-Meza M A 2012 *American Institute of Physics Conference Series* (*American Institute of Physics Conference Series* vol 1473) ed Ureña-López L A, Becerril-Bárcenas R and Linares-Romero R pp 74–84 (*Preprint* arXiv: 1210.6947)
[25] Martins C F and Salucci P 2007 *Mon. Not. R. Astron. Soc.* **381** 1103
[26] Randall L and Sundrum R 1999 *Phys. Rev. Lett.* **83** 3370–3373 (*Preprint* hep-ph/9905221)
[27] Randall L and Sundrum R 1999 *Phys. Rev. Lett.* **83** 4690–4693 (*Preprint* hep-th/9906064)
[28] Shiromizu T, Maeda K and Sasaki M 2000 *Phys. Rev. D* **62**(2) 024012 URL http://link.aps.org/doi/10.1103/PhysRevD.62.024012
[29] Maartens R 2000 *Phys. Rev. D* **62**(8) 084023
[30] Ovalle J and Linares F 2013 *Phys. Rev.* **D88** 104026
[31] Casadio R and Ovalle J 2012 *Physics Letters B* **715** 251 – 255 ISSN 0370-2693 URL http://www.sciencedirect.com/science/article/pii/S0370269312007812
[32] Ovalle J, Linares F, Pasqua A and Sotomayor A 2013 *Classical and Quantum Gravity* **30** 175019
[33] Germani C and Maartens R 2001 *Physical Review D* **64** 124010
[34] Garcia-Aspeitia M A 2014 *Rev. Mex. Fis.* **60** 205–209 (*Preprint* gr-qc:1306.1283)
[35] Langlois D, Maartens R, Sasaki M and Wands D 2001 *Physical Review D* **63** 084009
[36] García-Aspeitia M A and Ureña Lopez L A 2015 *Class. Quantum Grav.* **32** 025014 (*Preprint* gr-qc:1405.3932)
[37] Ovalle J and Linares F 2013 *Physical Review D* **88** 104026
[38] Perez-Lorenzana A 2005 *J. Phys. Conf. Ser.* **18** 224–269 (*Preprint* hep-ph/0503177)
[39] Ovalle J, Gergely L and Casadio R 2014 (*Preprint* 1405.0252)
[40] Garcia-Aspeitia M A, Reyes-Ibarra M J, Ortiz C, Lopez-Dominguez J and Hinojosa-Ruiz S 2014 (*Preprint* 1412.3496)
[41] Linares F X, García-Aspeitia M A and Ureña López L A 2015 *Phys. Rev. D* **92**(2) 024037 URL http://link.aps.org/doi/10.1103/PhysRevD.92.024037
[42] Casadio R and Germani C 2005 *Prog. Theor. Phys.* **114** 23–56 (*Preprint* hep-th/0407191)

IOP Conf. Series: Journal of Physics: Conf. Series **1010** (2018) 012006 doi:10.1088/1742-6596/1010/1/012006

[43] Kapner D, Cook T, Adelberger E, Gundlach J, Heckel B R *et al.* 2007 *Phys. Rev. Lett.* **98** 021101 (*Preprint* hep-ph/0611184)
[44] Alexeyev S, Rannu K, Dyadina P, Latosh B and Turyshev S 2015 (*Preprint* 1501.04217)
[45] Maartens R and Koyama K 2010 *Living Rev. Rel.* **13** 5 (*Preprint* 1004.3962)
[46] Sakstein J, Jain B and Vikram V 2014 *Int. J. Mod. Phys.* **D23** 1442002 (*Preprint* 1409.3708)
[47] Kudoh H, Tanaka T and Nakamura T 2003 *Phys. Rev.* **D68** 024035 (*Preprint* gr-qc/0301089)
[48] Cavaglia M 2003 *Int. J. Mod. Phys.* **A18** 1843–1882 (*Preprint* hep-ph/0210296)
[49] Holanda R, Silva J and Dahia F 2013 *Class. Quant. Grav.* **30** 205003 (*Preprint* 1304.4746)
[50] Barrow J D and Maartens R 2002 *Phys. Lett.* **B532** 153–158 (*Preprint* gr-qc/0108073)
[51] Brax P and van de Bruck C 2003 *Class. Quant. Grav.* **20** R201–R232 (*Preprint* hep-th/0303095)
[52] de Blok W, McGaugh S and Rubin V C 2001 *AJ* **122** 2396
[53] Persic M, Salucci P and Stel F 1996 *Mon. Not. Roy. Astron. Soc.* **281** 27
[54] Pavlidou V and Tomaras T N 2013 (*Preprint* 1310.1920)
[55] Navarro J F, Frenk C S and White S D M 1997 *Astrophys. J.* **490**
[56] Zhao H 1996 *Mon. Not. Roy. Astron. Soc.* **278** 488–496 (*Preprint* astro-ph/9509122)
[57] Spano M, Marcelin M, Amram P, Carignan C, Epinat B and Hernandez O 2008 *Mon. Not. Roy. Astron. Soc.* **383** 297–316 (*Preprint* 0710.1345)

IOP Conf. Series: Journal of Physics: Conf. Series **1010** (2018) 012007 doi:10.1088/1742-6596/1010/1/012007

Differential cross sections in a thick brane world scenario

Omar Pedraza

Área Académica de Matemáticas y Física, Universidad Autónoma del Estado de Hidalgo, Carretera Pachuca-Tulancingo Km. 4.5, C. P. 42184, Pachuca, México.

E-mail: omarp@uaeh.edu.mx

R. Arceo

Facultad de Ciencias en Física y Matemáticas, Universidad Autónoma de Chiapas, C. P. 29050, Tuxtla Gutiérrez, Chiapas, México.

E-mail: roberto.arceo@unach.mx

L. A. López

Área Académica de Matemáticas y Física, Universidad Autónoma del Estado de Hidalgo, Carretera Pachuca-Tulancingo Km. 4.5, C. P. 42184, Pachuca, México.

E-mail: lalopez@uaeh.edu.mx

V. E. Cerón

Área Académica de Matemáticas y Física, Universidad Autónoma del Estado de Hidalgo, Carretera Pachuca-Tulancingo Km. 4.5, C. P. 42184, Pachuca, México.

E-mail: vceron@uaeh.edu.mx

Abstract. The elastic differential cross section is calculated at low energies for the elements He and Ne using an effective 4D electromagnetic potential coming from the contribution of the massive Kaluza-Klein modes of the 5D vector field in a thick brane scenario. The length scale is adjusted in the potential to compare with known experimental data and to set bounds for the parameter of the model.

1. Introduction

The possibility that our space has more than three spatial dimensions has been attracting continuing interest for many years. Recently, the emphasis has shifted towards the so-called brane worlds scenarios. In the brane scenarios the ordinary matter are confined to a 3-brane embedded in higher-dimensional space-time [1, 2, 3].

In the brane world context, there are many models of thin branes [4, 5, 6] and thick branes [7]. Here, we use a recently proposed thick brane world configuration [8]. In this scenario the thick brane is modeled by an intriguing relation between the curvatures generated by the 5D and 4D cosmological constant without the inclusion of bulk scalar fields at all. An interesting feature of the thick brane world models is that singularities are not present at the position of the brane as other models [9, 10].

Within the framework of the brane world models, it has been seen that imply usually possible low energy consequences that should be either suppressed or else checked with experimental data. So the aim of this paper is to investigate some consequences of the extra dimensional theories at low energy. Specifically, we study the scattering of electrons by helium and neon in a thick brane scenario. Thus, for the sake of clarity, we first show the analysis of Ref. [11] to the case of the electromagnetic field. Finally we compare ours results for the scattering of electrons by atoms with known experimental data and to set bound on the scale of the scenarios.

The paper is organized as follows. In Sec. 2 we briefly describe the thick brane world setup, Sec. 3 is dedicated to obtain the static potential for an electromagnetic field. In Sec. 4 we discuss the elastic differential cross section for the scattering of electrons by atoms of different species. Finally Sec. 5 we focus on the discussion of our results.

2. Model

In this section we briefly describe the brane world used in the present work . The 5D action for the thick brane world model is

$$S = \frac{1}{8\pi G_5} \int d^5x\, \sqrt{-g}\, (R - 2\Lambda_5) \,, \tag{1}$$

here R is the $5D$ scalar curvature, Λ_5 is the bulk cosmological constant and G_5 is the $5D$ Newton constant. The action (1) provides a solution to Einstein equations with a cosmological constant in $5D$ that has the form

$$ds^2 = e^{2A(y)} \left(-dt^2 + a^2(t) \left[dx_1^2 + dx_2^2 + dx_3^2 \right] \right) + dy^2 \,, \tag{2}$$

where $e^{2A(y)}$ is the warp factor, $a(t)$ is the scale factor of the brane, t, x_1, x_2, x_3 denotes the $4D$ coordinates and y stands for the extra-dimensional coordinate.

The warp factor and the scale factor are given by

$$A(y) = \ln \left[\frac{H}{b} \cos(by) \right] \,, \tag{3}$$

$$a(t) = e^{Ht} \,. \tag{4}$$

Here H is the Hubble parameter and the $1/b$ parametrizes the thickness of the 3-brane. H and b are related to the 4D and 5D cosmological constant as

$$3H^2 = \Lambda_4 \,, \tag{5}$$

$$\Lambda_5 = 6b^2, \tag{6}$$

where Λ_4 is obtained upon integration of the fifth dimension. In this model, the 3-brane is located at $y = 0$ and the range of the fifth dimension is $-|\pi/2b| < y < |\pi/2b|$.

For future purposes, it turns out to be convenient to perform the change of variable $dz = e^{-A(y)}dy$. One gets the expression for z

$$z(y) = \frac{2}{H} arctanh \left[\tan \left(\frac{by}{2} \right) \right] \,. \tag{7}$$

Due to this transformation, the warp factor adopts the form

$$A(z) = \ln \left[\frac{H}{b} sech\, (Hz) \right] \,, \tag{8}$$

and the metric can be written as

$$ds^2 = \frac{H}{b} sech\,(Hz)\left[-dt^2 + e^{2Ht}\left(dx_1^2 + dx_2^2 + dx_3^2\right) + dz^2\right]. \tag{9}$$

In this scenario the effective 4D Planck mass is

$$M_{Pl}^2 = \frac{\pi}{2}\frac{M_*^3 H^2}{b^3}, \tag{10}$$

with M_* the 5D Planck mass. By assigning $H \approx 10^{-60} M_{\mathbf{Pl}}$ and setting $M_* \approx 10^{-55} M_{\mathbf{Pl}}$, one gets the correct 4D gravitational coupling on the thick brane. On the other hand, in order to generate the desired mass hierarchy, in this model we need that the ratio between the Hubble parameter and the compactification scale is

$$\frac{H}{b} \approx 10^{-5}. \tag{11}$$

Of course this reformulation of the hierarchy problem raises the questions about the stability of the brane separation.

3. Effective four-dimensional electrostatic potential

In this section, we show the analysis to calculate corrections to Coulomb's law developed in Ref. [11]. It is known that the charged fermions in quantum electrodynamics interact with one another by exchanging a photon. Hence the interaction between fermions and a gauge field can be written as

$$S_I = \int d^4x dz\, \sqrt{-g}(-e_5)\bar{\Psi}(x,z)\Gamma^M \Psi(x,z) A_M(x,z), \tag{12}$$

where the e_5 is a 5D coupling constant. After the dimensional reduction (see [11] for more details), we obtain the following coupling of fermion zero mode and photon Kaluza-Klein modes

$$\begin{aligned} S_I &= (-e_5)\sum_n \int dz\; e^{-A/2}\, \rho_n(z) L_0^2(z) \int d^4x \sqrt{-\hat{g}}\; \bar{\psi}_0(x)\gamma^\mu a_\mu^{(n)}(x)\psi_0(x), \\ &= \int d^4x\, \sqrt{-\hat{g}}\, \Big\{ - e\bar{\psi}_0(x)\gamma^\mu a_\mu^{(0)}(x)\psi_0(x) - \sum_{n\neq 0} \epsilon_n \bar{\psi}_0(x)\gamma^\mu a_\mu^{(n)}(x)\psi_0(x) \Big\}. \end{aligned} \tag{13}$$

Here, e is the usual 4D charge of fermion zero mode, given by

$$e = e_5 \int dz\, e^{-A/2} \rho_0(z) L_0^2(z) = e_5 \sqrt{\frac{b}{\pi}}, \tag{14}$$

and ϵ_n's are 4D effective couplings

$$\epsilon_n = e_5 \int dz\, e^{-A/2} \rho_n(z) L_0^2(z) = e\sqrt{\frac{\pi}{b}} \int dz\, e^{-A/2} \rho_n(z) L_0^2(z). \tag{15}$$

$L_0(z)$ is the zero mode for the 5D fermion, $\rho_0(z)$ and $\rho_n(z)$ are the Kaluza-Klein modes for the 5D vector field, whose wave functions are given by

$$L_0(z) = \left[\frac{H\Gamma\left(\frac{b+2M}{2b}\right)}{\sqrt{\pi}\,\Gamma\left(\frac{M}{b}\right)}\right] sech^{\frac{M}{b}}(Hz), \tag{16}$$

$$\rho_0(z) = \sqrt{\frac{H}{\pi} sech(Hz)}, \tag{17}$$

$$\rho_n(z) = C_+ P_{1/2}^{+i\beta}\left[\tanh(Hz)\right] + C_- P_{1/2}^{-i\beta}\left[\tanh(Hz)\right], \tag{18}$$

with $\Gamma(z)$ the Gamma function and $P^{\mu}_{1/2}(z)$ y $Q^{\mu}_{1/2}(z)$ are associated Legendre functions of first and second kind, respectively, degree $\nu = 1/2$ and purely imaginary order $\mu = i\beta = i\sqrt{m^2/H^2 - 1/4}$. The constants C_+ y C_- are expressed as follows

$$C_{\pm}(\beta) = \frac{\Gamma\left(1 \mp i\beta\right)}{\sqrt{2\pi}} = \frac{|\Gamma\left(1 + i\beta\right)|}{\sqrt{2\pi}} \,. \tag{19}$$

$\sum_n$ implies summations or integrations (or both) with respect to n, depending on the respective discrete or continuous character of the $a^{(n)}_{\mu}(x)$ and $\epsilon_n(z)$, where latter is determined by $\rho_n(z)$.

In the non-relativistic limit, the potential between two charged fermion zero modes can be found by the Kaluza-Klein photon exchange process as

$$v(r) = \frac{e^2}{4\pi r} + \int_{H/2}^{\infty} dm\, \frac{\epsilon_n^2}{4\pi r} e^{-mr}\,, \tag{20}$$

$$= \frac{e^2}{4\pi r}\left[1 + \frac{\pi}{b}\int_{H/2}^{\infty} dm\, e^{-mr}\left(\int dz\, e^{-A/2}\rho_n(z) L_0^2(z)\right)^2\right], \tag{21}$$

where r is the spatial separation in 4D between charged fermions. The first term of (20) represents the contribution of massless photon, while remaining terms come from its massive Kaluza-Klein modes.

Substituting the expressions (16)-(18) in Eq. (21), we obtain

$$v(r) = \frac{e^2}{4\pi r}\left[1 + \frac{\pi}{H}\left[\frac{\Gamma\left(\frac{2M+b}{2b}\right)\Gamma\left(\frac{4M-b}{4b}\right)}{\Gamma\left(\frac{M}{b}\right)\;\Gamma\left(\frac{4M+b}{4b}\right)}\right]^2 \int_{H/2}^{\infty} dm\, e^{-mr}\left|\sum_{\pm} C_{\pm}\left(\beta\right) P^{\pm i\beta}_{1/2}\left(0\right)\right|^2\right], \tag{22}$$

where we have used the following relation

$$\delta(z) = \lim_{H\to\infty} \frac{H\Gamma\left(\frac{4M+b}{4b}\right)}{\sqrt{\pi}\Gamma\left(\frac{4M-b}{4b}\right)} sech^{\frac{4M-b}{2b}}(Hz)\,, \quad 4M > b\,. \tag{23}$$

Taking into account Eq. (19) as well as the following reaction

$$P^{\mu}_{\nu}(0) = \frac{2^{\mu}\sqrt{\pi}}{\Gamma\left(\frac{1-\nu-\mu}{2}\right)\Gamma\left(1 + \frac{\nu-\mu}{2}\right)}\,. \tag{24}$$

The expression (22) can be rewritten as

$$v(r) = \frac{e^2}{4\pi r}\left[1 + \frac{2\pi}{H}\left[\frac{\Gamma\left(\frac{2M+b}{2b}\right)\Gamma\left(\frac{4M-b}{4b}\right)}{\Gamma\left(\frac{M}{b}\right)\;\Gamma\left(\frac{4M+b}{4b}\right)}\right]^2 \int_{H/2}^{\infty} dm\, e^{-mr}\left|\frac{\Gamma\left(1 + i\beta\right))}{\Gamma\left(\frac{1}{4} - \frac{i\beta}{2}\right)\Gamma\left(\frac{5}{4} - \frac{i\beta}{2}\right)}\right|^2\right]. \tag{25}$$

To perform the integral in Eq. (25), we do an expansion in the integrand with respect to $H/2$, i.e. around $\beta = 0$. So, the leading-order contribution for the potential $v(r)$ is

$$v(r) \approx \frac{e^2}{4\pi r}\left[1 + 2\pi\left[\frac{\Gamma\left(\frac{2M+b}{2b}\right)\Gamma\left(\frac{4M-b}{4b}\right)}{\Gamma\left(\frac{M}{b}\right)\;\Gamma\left(\frac{4M+b}{4b}\right)}\right]^2 \frac{1}{\left|\Gamma\left(\frac{1}{4}\right)\Gamma\left(\frac{5}{4}\right)\right|^2}\frac{e^{-Hr/2}}{Hr} + \ldots\right]. \tag{26}$$

The first term in (26) is the usual Coulomb interaction potential which coming from the exchange of the zero mode of the photon and the second term has a Yukawa potential form because it is mediated by an infinite tower of massive Kaluza-Klein vector modes. Also the corresponding correction to 4D effective potential depends on the 5D paramters of the massless zero mode fermion field M and b.

In the following section we study the scattering using this effective potential.

IOP Conf. Series: Journal of Physics: Conf. Series **1010** (2018) 012007 doi:10.1088/1742-6596/1010/1/012007

4. Scattering by helium and neon atoms

In this section we study the collision of a particle of charge ze and mass m with a neutral atom of atomic number Z. Within the Born approximation the differential scattering cross section for elastic processes is given by

$$\frac{d\sigma}{d\Omega} = \left| -\frac{m}{2\pi\hbar^2} \int e^{i\vec{q}\cdot\vec{r}}\, V(\vec{r})\, d^3\vec{r} \right|^2 , \tag{27}$$

where $\vec{r}$ is the position vector of the incident particle and the momentum transfer $\vec{q}$ is given in terms of θ and the incident particle momentum $\vec{k}$ as

$$|\vec{q}| = 2\left|\vec{k}\right| \sin\left(\theta/2\right) . \tag{28}$$

The electrostatic potential $V(r)$ considers the effective interaction of the incident particle with the constituents of the neutral atom. Considering the nucleus as a point charge, $V(r)$ can be calculated in terms of the atomic electron density $\rho(\vec{r})$ as

$$V\left(r\right) = ze\left[Zev\left(\vec{r}\right) + e\int \rho\left(\vec{r}\right) v\left(\left|\vec{r}-\vec{r}\,'\right|\right)\, d^3\vec{r}\,'\right] . \tag{29}$$

$v(\vec{r})$ is given by (26). Here we are ignoring all effects of symmetry and spin for the time being. For neutral atoms, the density satisfies

$$\int \rho\left(\vec{r}\right) d^3\vec{r} = Z . \tag{30}$$

Thus introducing Eq. (29) in (27) and making the following change of variable $\vec{R} = \vec{r} - \vec{r}\,'$, thus we have

$$\begin{aligned} \frac{d\sigma}{d\Omega} &= \left| -\frac{me^2}{2\pi^{-}\hbar}\left[zZ\int e^{i\vec{q}\cdot\vec{r}} v\left(\vec{r}\right) d^3\vec{r} - z\int e^{i\vec{q}\cdot\vec{R}} v\left(\vec{R}\right) d^3\vec{R} \int \rho\left(\vec{r}\,'\right) e^{i\vec{q}\cdot\vec{r}\,'} d^3\vec{r}\,' \right]\right|^2 , \\ &= \frac{m^2e^4z^2}{4\pi^{2-}\hbar}\left[Z - F(\vec{q})\right]^2 \left|\int e^{i\vec{q}\cdot\vec{r}} v\left(\vec{r}\right) d^3\vec{r}\right|^2 , \end{aligned} \tag{31}$$

where $F(\vec{q})$ is called the form factor and it is defined as

$$F(\vec{q}) = \int \rho\left(\vec{r}\right) e^{i\vec{q}\cdot\vec{r}} d^3\vec{r}. \tag{32}$$

When the potential $v(r)$ is spherically symmetric, the angular integration in (31) can be performed to give

$$\frac{d\sigma}{d\Omega} = \frac{4m^2e^4z^2}{{}^{-}\hbar}\left[Z - F(\vec{q})\right]^2 \left|\int_0^\infty \frac{\sin(qr)}{qr} v\left(r\right) r^2\, dr\right|^2 . \tag{33}$$

Considering the form of $v(r)$ and using the following relation (see [12])

$$\int_0^\infty \frac{e^{-pr}\sin(qr)}{r}\, dr = \arctan\left(\frac{q}{p}\right), \quad p > 0 . \tag{34}$$

The differential scattering cross section reduces to

$$\frac{d\sigma}{d\Omega} = \left(\frac{2me^2z}{{}^{-}\hbar}\right)^2 \left[Z - F(\vec{q})\right]^2 \left[\frac{1}{q} + \left(\frac{\Gamma\left(\frac{2M+b}{2b}\right)\Gamma\left(\frac{4M-b}{4b}\right)}{\Gamma\left(\frac{M}{b}\right)\Gamma\left(\frac{4M+b}{4b}\right)}\right)^2 \frac{1}{\left[\Gamma\left(\frac{1}{4}\right)\Gamma\left(\frac{5}{4}\right)\right]^2} \frac{2\pi}{qH} \arctan\left(\frac{2q}{H}\right)\right]^2 . \tag{35}$$

For many-electrons neutral atoms, exact solutions for the atomic form factor are not obtainable, so a variety of approximations in the literature have been used. Among these approaches, an analytical approximation developed in [3] for the form factor was proposed by using the statistical method Thomas-Fermi. Thus the form factor is given by

$$F(q) = Z\sum_{i=1}^{3} \frac{A_i\alpha_i^2}{\alpha_i^2 + a_0^2q^2}, \tag{36}$$

with $a_0 = \hbar^2/me^2$ being the Bohr radius and the A_i and α_i parameters are tabulated in [13] for Z values from 1 to 92. The parameters in (36) for helium and neon atoms are given in Table 1. So, the differential scattering cross section for elastic processes becomes

$$\frac{d\sigma}{d\Omega} = \left(\frac{2zZ}{a_0q^2}\right)^2 \left[1 - \sum_{i=1}^{3} \frac{A_i\alpha_i^2}{\alpha_i^2 + a_0^2q^2}\right]^2 \left[1 + \left(\frac{\Gamma\left(\frac{2M+b}{2b}\right)\Gamma\left(\frac{4M-b}{4b}\right)}{\Gamma\left(\frac{M}{b}\right)\Gamma\left(\frac{4M+b}{4b}\right)}\right)^2 \frac{1}{\left[\Gamma\left(\frac{1}{4}\right)\Gamma\left(\frac{5}{4}\right)\right]^2} \frac{2\pi}{H} \arctan\left(\frac{2q}{H}\right)\right]^2. \tag{37}$$

Table 1. Parameters of the form factor for helium and neon atoms.

Element	A_1	A_2	A_3	α_1	α_2	α_3
He	−0.2259	1.2259	0	5.5272	2.3992	0
Ne	0.0188	0.9812	0	34.999	2.5662	0

For incident electrons, we set $z = -1$ and for helium and neon atoms, we have $Z = 2$ and $Z = 10$ respectively. To complete the analysis, we evaluate numerically (37) and we compare the theoretical results with experimental data [14] for differential cross section for elastic scattering of electrons by helium and neon atoms. This comparison is made explicit in Figs. 1 and 2 .

Thus, in Figure 1 the elastic differential cross section is analyzed for electrons by helium . We evaluate the elastic cross section (37) at 2000 eV and 3000 eV, varying the parameter M to adjust the experimental points. A best agreement between theoretical differential cross section and experimental data is attained when $M \sim 10^{-43}$ eV.

In the Fig. 2 it is shown that there are discrepancies between theoretical and experimental results for scattering of electrons off neon at small angles for the point line, while for the dashed line, the qualitative agreement is not good.

5. Conclusions

In this work we have explored the corrections to Coulomb's law coming from massive Kaluza-Klein vector modes in a thick brane scenario. For this purpose, we included the analyses accomplished in reference for the static potential in the low-energy regime. The corresponding corrections to Coulomb's law decays exponentially and depend on the 5D parameters of the zero massless fermion field M and b.

To this end, we have studied numerically this problem making a comparison between the total cross section for electrons scattered by different atoms modified by incorporation of the corrections to Coulomb's law in a thick brane scenario and the curve experimental data for each case (see Figures 1, 2). This comparison yields the bound of $M = 3 \times 10^{-45}$ eV.

Finally, we mention that in the low energy regime we focused on in this work there are other possible directions which can be pursued, so we wonder whether the same results or behavior holds for the gravitational case.

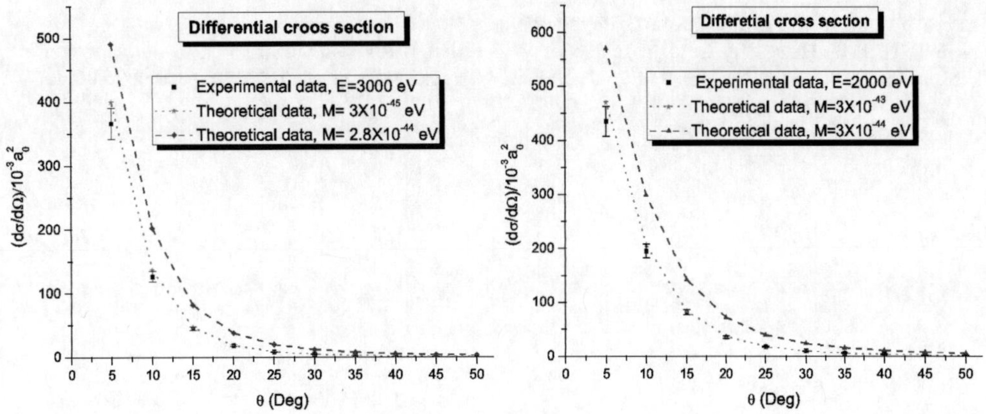

Figure 1. Comparison of theoretical and experimental elastic differential cross section for electrons off helium. We set $H = 10^{-60} M_{Pl}$ and $b = 10^{-55} M_{Pl}$.

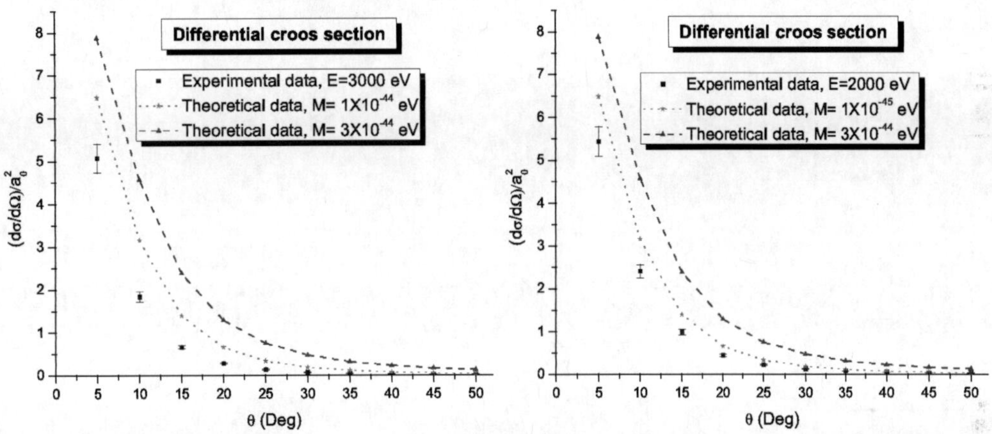

Figure 2. Comparison of theoretical and experimental elastic differential cross section for electrons off neon. We set $H = 10^{-60} M_{Pl}$ and $b = 10^{-55} M_{Pl}$ again.

References

[1] Rubakov V and Shaposhnikov M 1983 *Physics Letters B* **125** 136 – 138 ISSN 0370-2693 URL `http://www.sciencedirect.com/science/article/pii/0370269383912534`
[2] Arkani-Hamed N, Dimopoulos S and Dvali G R 1998 *Phys. Lett.* **B429** 263–272 (*Preprint* `hep-ph/9803315`)
[3] Antoniadis I, Arkani-Hamed N, Dimopoulos S and Dvali G R 1998 *Phys. Lett.* **B436** 257–263 (*Preprint* `hep-ph/9804398`)
[4] Gogberashvili M 2002 *Int. J. Mod. Phys.* **D11** 1635–1638 (*Preprint* `hep-ph/9812296`)
[5] Randall L and Sundrum R 1999 *Phys. Rev. Lett.* **83** 3370–3373 (*Preprint* `hep-ph/9905221`)
[6] Gogberashvili M 1999 *Mod. Phys. Lett.* **A14** 2025–2032 (*Preprint* `hep-ph/9904383`)
[7] Dzhunushaliev V, Folomeev V and Minamitsuji M 2010 *Rept. Prog. Phys.* **73** 066901 (*Preprint* `0904.1775`)
[8] Herrera-Aguilar A, Malagon-Morejon D and Mora-Luna R R 2010 *JHEP* **11** 015 (*Preprint* `1009.1684`)
[9] Maartens R and Koyama K 2010 *Living Rev. Rel.* **13** 5 (*Preprint* `1004.3962`)
[10] Shifman M 2010 *Int. J. Mod. Phys.* **A25** 199–225 (*Preprint* `0907.3074`)
[11] Guo H, Herrera-Aguilar A, Liu Y X, Malagon-Morejon D and Mora-Luna R R 2013 *Phys. Rev.* **D87** 095011 (*Preprint* `1103.2430`)
[12] Gradshteyn I S, Ryzhik I M, Jeffrey A, Zwillinger D and Scripta Technica I 2007 *Table of integrals, series, and products* (Amsterdam, Boston, Paris, et al.: Elsevier) ISBN 0-12-373637-4 trad. de : Tablitsy integralov, summ, riadov i proizvedeni URL `http://opac.inria.fr/record=b1122421`
[13] Salvat F, Martínez J D, Mayol R and Parellada J 1987 *Phys. Rev. A* **36**(2) 467–474 URL

http://link.aps.org/doi/10.1103/PhysRevA.36.467
[14] Jansen R H J, de Heer F J, Luyken H J, van Wingerden B and Blaauw H J 1976 *Journal of Physics B: Atomic and Molecular Physics* **9** 185 URL http://stacks.iop.org/0022-3700/9/i=2/a=009

A premier analysis of supersymmetric closed string tachyon cosmology

V Vázquez-Báez[1] and C Ramírez[2]

[1]Benemérita Universidad Autónoma de Puebla, Facultad de Ingeniería, Blvd. Valsequillo esq. Av. San Claudio s/n, Ciudad Universitaria, Edificio 108-C, 72570 Puebla, México
[2]Benemérita Universidad Autónoma de Puebla, Facultad de Ciencias Físico Matemáticas, P.O. Box 165, 72000 Puebla, México

E-mail: manuel.vazquez@correo.buap.mx

Abstract. From a previously found worldline supersymmetric formulation for the effective action of the closed string tachyon in a FRW background, the Hamiltonian of the theory is constructed, by means of the Dirac procedure, and written in a quantum version. Using the supersymmetry algebra we are able to find solutions to the Wheeler-DeWitt equation via a more simple set of first order differential equations. Finally, for the $k = 0$ case, we compute the expectation value of the scale factor with a suitably potential also favored in the present literature. We give some interpretations of the results and state future work lines on this matter.

1. Introduction

Tachyons correspond to the lowest mode of string theory, which for the lack of knowledge on how to handle unstable configurations were ignored for many years. With the inclusion of supersymmetry they can be consistently eliminated from the spectrum by the GSO truncation. Some years ago it has been seen that the evolution of these tachyonic instabilities can be described by the condensation of the tachyonic modes. This was first perfomed in the simpler case of open strings, resumed by the well known Sen conjectures [1]. Whereas for closed strings the situation is more complicated, because it involves the structure of space-time. A key element for the development of the present work is the interesting fact that closed string tachyons, which are nonsupersymmetric in the target space, can have worldsheet supersymmetry [2]. Supersymmetric cosmology has been studied in a variety of different schemes, we refer the interested reader to the well known books by D'Eath and Moniz for a review [3, 4]. In [5] a Lagrangian of supersymmetric tachyons in the framework of a FRW background was given in a worldline superspace where, the time variable is extended to the superspace of supersymmetry [6, 7], this is done considering the covariant formulation of one-dimensional supergravity of the so called 'new' Θ variables [8, 9], which allows in a straightforward way to write supergravity invariant actions. The present work is performed along such approach, we depart from the Hamiltonian given in [5], then we give a particular, and normalizable, solution to the Wheeler-DeWitt equation via the superalgebra constraints, which act as square roots of the WDW equation. We compute the scale factor expectation value as customary. Plots of the results are given, follow up by some conclusions and final remarks on future work.

IOP Conf. Series: Journal of Physics: Conf. Series **1010** (2018) 012008 doi:10.1088/1742-6596/1010/1/012008

2. Closed String Tachyon Effective Action

The bosonic closed string tachyon effective action is given according to [10] as

$$S = \frac{1}{2\kappa_D^2}\int \sqrt{-g}e^{-2\phi}\left[R + 4(\partial\phi)^2 - (\partial T)^2 - 2V(T)\right]d^Dx, \tag{1}$$

where T is the closed string tachyon field, $V(T)$ is the tachyon potential and ϕ is the dilaton field. We write this action in the Einstein frame, by means of $g_{\mu\nu}^{string} = e^{\phi}g_{\mu\nu}^{Einstein}$, which is more suitable for our cosmological approach, in this frame and for a 4−dimensional FRW metric the action takes the form

$$S = \int\left[-\frac{3\dot{a}^2 a}{\kappa^2 N} + \frac{3Nka}{\kappa^2} + \frac{a^3}{\kappa^2 N}\dot{\phi}^2 + \frac{a^3}{2\kappa^2 N}\dot{T}^2 - \frac{a^3 N e^{2\phi}V(T)}{\kappa^2}\right]dt, \tag{2}$$

where N is the lapse function and a is the scale factor. The natural invariance of this Lagrangian under time reparametrizations is extended to supersymmetry introducing a Grassmann superspace associated to the bosonic time coordinate t (see Tkach et al. in [6, 7]).

3. Supersymetric Closed String Tachyon Model

Superspace is the natural framework for a geometrical formulation of supersymmetry and supergravity [11], it extends spacetime by anticommuting Grassmann variables, $x^m \to (x^m, \theta^\mu)$, and the field content of the superfields is given by the Grassmann power expansion in the anticommuting variables $\phi(z) = \sum_n 1/n!\,\theta^{\mu_1}\cdots\theta^{\mu_n}\phi_{\mu_1\cdots\mu_n}(x)$.

The supersymmetric cosmological model for the action (2) is obtained upon such extension of the time coordinate into a supermultiplet $t \to (t, \Theta, \bar{\Theta})$. Due to this generalization the fields of the theory also are generalized as superfields, the power expansion mentioned above is given explicitly by

$$\begin{aligned}\mathcal{A}\left(t,\Theta,\bar{\Theta}\right) &= a(t) + i\Theta\bar{\lambda}(t) + i\bar{\Theta}\lambda(t) + B(t)\Theta\bar{\Theta},\\ \mathcal{T}\left(t,\Theta,\bar{\Theta}\right) &= T(t) + i\Theta\bar{\eta}(t) + i\bar{\Theta}\eta(t) + G(t)\Theta\bar{\Theta},\\ \Phi\left(t,\Theta,\bar{\Theta}\right) &= \phi(t) + i\Theta\bar{\chi}(t) + i\bar{\Theta}\chi(t) + F(t)\Theta\bar{\Theta},\end{aligned} \tag{3}$$

where, $\mathcal{A}$, $\mathcal{T}$ and Φ are the superfields corresponding to a, T and ϕ, respectively.

The supersymmetric generalization of the action is

$$S = S_{Rsusy} + S_{Msusy}, \tag{4}$$

where, S_{Rsusy} is the cosmological supersymmetric generalization of the free FRW model

$$S_{Rsusy} = \frac{1}{\kappa^2}\int\left(3\mathcal{E}\mathcal{A}\nabla_{\bar{\Theta}}\mathcal{A}\nabla_{\Theta}\mathcal{A} - 3\sqrt{k}\mathcal{E}\mathcal{A}^2\right)d\Theta d\bar{\Theta}dt, \tag{5}$$

and the supersymmetric matter term is

$$S_{Msusy} = \frac{1}{\kappa^2}\int\left[-\mathcal{E}\mathcal{A}^3\nabla_{\bar{\Theta}}\Phi\nabla_{\Theta}\Phi - \frac{1}{2}\mathcal{E}\mathcal{A}^3\nabla_{\bar{\Theta}}\mathcal{T}\nabla_{\Theta}\mathcal{T} + \mathcal{E}\mathcal{A}^3 W(\Phi,\mathcal{T})\right]d\Theta d\bar{\Theta}dt, \tag{6}$$

where $W(\Phi,\mathcal{T})$ is the superpotential, and $\mathcal{E}$ is an invariant density playing the role of a supersymmetric $\sqrt{-g}$ in the action and given by

$$\mathcal{E} = -e - \frac{i}{2}(\Theta\bar{\Psi} + \bar{\Theta}\psi), \tag{7}$$

we refer the interested reader to [5, 8, 9] for full details of its derivation.

The superpotential is also written as a power series expansion, given as $W(\Phi, \mathcal{T}) = W(\phi, T) + \frac{\partial W}{\partial \phi}(\Phi - \phi) + \frac{\partial W}{\partial T}(\mathcal{T} - T) + \frac{1}{2}\frac{\partial^2 W}{\partial T^2}(\mathcal{T} - T)^2 + \frac{1}{2}\frac{\partial^2 W}{\partial \phi^2}(\Phi - \phi)^2 + \frac{\partial^2 W}{\partial T \partial \phi}(\mathcal{T} - T)(\Phi - \phi)$. This expansion is finite because the terms $(\mathcal{T} - T)$ and $(\Phi - \phi)$ are purely grassmannian and nilpotency holds for these kind of variables.

The total Lagrangian is found integrating over the Grassmann parameters Θ and $\bar{\Theta}$. After we perform the variation of the Lagrangian with respect to the fields B, F and G they can be eliminated from the Lagrangian, as usual their equations of motion are algebraic constraints, that is B, F and G are auxiliary fields. Upon solving for the auxiliary fields and making the further rescalings for convinience $\lambda \to \kappa a^{-1/2}\lambda$, $\bar{\lambda} \to \kappa a^{-1/2}\bar{\lambda}$, $\eta \to \kappa a^{-3/2}\eta$, $\bar{\eta} \to \kappa a^{-3/2}\bar{\eta}$, $\chi \to \kappa a^{-3/2}\chi$, $\bar{\chi} \to \kappa a^{-3/2}\bar{\chi}$, the Lagrangian is written as

$$
\begin{aligned}
L = & -\frac{3a\dot{a}^2}{e\kappa^2} + \frac{3\sqrt{a}\dot{a}}{e\kappa}\left(\psi\lambda - \bar{\psi}\bar{\lambda}\right) + \frac{3eka}{\kappa^2} + \frac{\dot{T}^2 a^3}{2e\kappa^2} - \frac{\sqrt{a^3}\dot{T}}{2e\kappa}\left(\psi\eta - \bar{\psi}\bar{\eta}\right) + \frac{3i\dot{T}}{2}\left(\lambda\bar{\eta} + \bar{\lambda}\eta\right) \\
& + \frac{\dot{\phi}^2 a^3}{e\kappa^2} - \frac{\sqrt{a^3}\dot{\phi}}{e\kappa}\left(\psi\chi - \bar{\psi}\bar{\chi}\right) + 3i\dot{\phi}\left(\lambda\bar{\chi} + \bar{\lambda}\chi\right) + 3i\left(\lambda\dot{\bar{\lambda}} + \bar{\lambda}\dot{\lambda}\right) - \frac{i}{2}\left(\eta\dot{\bar{\eta}} + \bar{\eta}\dot{\eta}\right) \\
& - i\left(\chi\dot{\bar{\chi}} + \bar{\chi}\dot{\chi}\right) + \frac{3e\sqrt{k}\lambda\bar{\lambda}}{a} - \frac{3e\sqrt{k}\eta\bar{\eta}}{2a} - \frac{3e\sqrt{k}\chi\bar{\chi}}{a} + \frac{3i\sqrt{ak}}{\kappa}\left(\psi\lambda + \bar{\psi}\bar{\lambda}\right) \\
& + \frac{3eW^2a^3}{4\kappa^2} - \frac{3e\sqrt{k}a^2W}{\kappa^2} - \frac{eW_T^2a^3}{2\kappa^2} - \frac{eW_\phi^2a^3}{4\kappa^2} - \frac{9}{2}eW\lambda\bar{\lambda} + \frac{3}{4}eW\eta\bar{\eta} + \frac{3}{2}eW\chi\bar{\chi} \\
& - \frac{3ia^{3/2}W}{2\kappa}\left(\psi\lambda + \bar{\psi}\bar{\lambda}\right) - \frac{i\sqrt{a^3}W_T}{2\kappa}\left(\psi\eta + \bar{\psi}\bar{\eta}\right) + \frac{3eW_T}{2}\left(\bar{\lambda}\eta - \lambda\bar{\eta}\right) - \frac{i\sqrt{a^3}W_\phi}{2\kappa}\left(\psi\chi + \bar{\psi}\bar{\chi}\right) \\
& + \frac{3eW_\phi}{2}\left(\bar{\lambda}\chi - \lambda\bar{\chi}\right) - eW_{TT}\eta\bar{\eta} + eW_{T\phi}\left(\bar{\chi}\eta - \chi\bar{\eta}\right) - eW_{\phi\phi}\chi\bar{\chi} + \frac{3e\kappa^2}{4a^3}\eta\bar{\eta}\chi\bar{\chi} - \frac{3}{2e}\psi\bar{\psi}\lambda\bar{\lambda} \\
& + \frac{1}{2e}\psi\bar{\psi}\chi\bar{\chi} + \frac{1}{4e}\psi\bar{\psi}\eta\bar{\eta},
\end{aligned}
$$

where the subscripts in W denote partial differentiation with respect to ϕ and T, respectively.

As usual, when we compute the canonical momenta, in order to obtain the Hamiltonian of the theory, we have the appearance of fermionic constraints, which are second class. Thus, we must follow Dirac formalism to obtain the dynamics of the system. Using the standard definition for the Hamiltonian and according to Dirac's prodecure we can write

$$
H = NH_0 + \frac{1}{2}\psi S - \frac{1}{2}\bar{\psi}\bar{S}, \tag{8}
$$

where

$$
\begin{aligned}
H_0 = & -\tfrac{\kappa^2\pi_a^2}{12a} + \tfrac{\kappa^2\pi_T^2}{2a^3} - \tfrac{3i\kappa^2\pi_T}{2a^3}\left(\lambda\bar{\eta} + \bar{\lambda}\eta\right) + \tfrac{\kappa^2\pi_\phi^2}{4a^3} - \tfrac{3i\kappa^2\pi_\phi}{2a^3}\left(\lambda\bar{\chi} + \bar{\lambda}\chi\right) - \tfrac{3a^3}{4\kappa^2}W^2 \\
& + \tfrac{3\sqrt{k}a^2}{\kappa^2}W - \tfrac{3ka}{\kappa^2} + \tfrac{a^3}{2\kappa^2}W_T^2 + \tfrac{a^3}{4\kappa^2}W_\phi^2 + \tfrac{9}{2}W\lambda\bar{\lambda} - \tfrac{3}{4}W\eta\bar{\eta} - \tfrac{3}{2}W\chi\bar{\chi} \\
& + \tfrac{3}{2}W_T\left(\lambda\bar{\eta} - \bar{\lambda}\eta\right) + \tfrac{3}{2}W_\phi\left(\lambda\bar{\chi} - \bar{\lambda}\chi\right) + W_{TT}\eta\bar{\eta} + W_{T\phi}\left(\chi\bar{\eta} - \bar{\chi}\eta\right) \\
& + W_{\phi\phi}\chi\bar{\chi} - \tfrac{3\sqrt{k}}{a}\lambda\bar{\lambda} + \tfrac{3\sqrt{k}}{2a}\eta\bar{\eta} + \tfrac{3\sqrt{k}}{a}\chi\bar{\chi} - \tfrac{9\kappa^2}{2a^3}\lambda\bar{\lambda}\chi\bar{\chi} - \tfrac{9\kappa^2}{4a^3}\lambda\bar{\lambda}\eta\bar{\eta} - \tfrac{3\kappa^2}{4a^3}\eta\bar{\eta}\chi\bar{\chi},
\end{aligned} \tag{9}
$$

$$
\begin{aligned}
S = & \tfrac{\kappa\pi_a}{\sqrt{a}}\lambda + \tfrac{\kappa\pi_T}{\sqrt{a^3}}\eta + \tfrac{\kappa\pi_\phi}{\sqrt{a^3}}\chi - \tfrac{6i\sqrt{ak}}{\kappa}\lambda + \tfrac{3i\sqrt{a^3}}{\kappa}W\lambda \\
& + \tfrac{i\sqrt{a^3}}{\kappa}W_T\eta + \tfrac{i\sqrt{a^3}}{\kappa}W_\phi\chi + \tfrac{3i\kappa}{2a^{3/2}}\lambda\eta\bar{\eta} + \tfrac{3i\kappa}{a^{3/2}}\lambda\chi\bar{\chi},
\end{aligned} \tag{10}
$$

$$
\begin{aligned}
\bar{S} = & \tfrac{\kappa\pi_a}{\sqrt{a}}\bar{\lambda} + \tfrac{\kappa\pi_T}{\sqrt{a^3}}\bar{\eta} + \tfrac{\kappa\pi_\phi}{\sqrt{a^3}}\bar{\chi} + \tfrac{6i\sqrt{ak}}{\kappa}\bar{\lambda} - \tfrac{3i\sqrt{a^3}}{\kappa}W\bar{\lambda} \\
& - \tfrac{i\sqrt{a^3}}{\kappa}W_T\bar{\eta} - \tfrac{i\sqrt{a^3}}{\kappa}W_\phi\bar{\chi} - \tfrac{3i\kappa}{2a^{3/2}}\bar{\lambda}\eta\bar{\eta} - \tfrac{3i\kappa}{a^{3/2}}\bar{\lambda}\chi\bar{\chi},
\end{aligned} \tag{11}
$$

are the set of first class constraints satisfying the Dirac algebra $\{S, \bar{S}\}_D = 2H_0$, $\{H_0, S\}_D = \{H_0, \bar{S}\}_D = 0$.

IOP Conf. Series: Journal of Physics: Conf. Series **1010** (2018) 012008 doi:10.1088/1742-6596/1010/1/012008

4. Wave function of the universe

In the quantum version the constraints (9)-(11) become conditions on the wave function of the universe, i.e, $H\Psi = \bar{S}\Psi = S\Psi = 0$, which due to the operator superalgebra imply that any state wave function, Ψ, satisfying $\bar{S}\Psi = S\Psi = 0$ is also solution of the Wheeler-DeWitt equation, $H\Psi = 0$. We have for the quantum cosntraints:

$$S = \frac{\kappa\pi_a}{\sqrt{a}}\lambda + \frac{\kappa\pi_\phi}{\sqrt{a^3}}\chi + \frac{\kappa\pi_T}{\sqrt{a^3}}\eta + \frac{3i\sqrt{a^3}W}{\kappa}\lambda + \frac{i\sqrt{a^3}W_T}{\kappa}\eta + \frac{i\sqrt{a^3}W_\phi}{\kappa}\chi + \frac{3i\kappa}{4a^{3/2}}\lambda\left[\eta,\bar{\eta}\right] + \frac{3i\kappa}{2a^{3/2}}\lambda\left[\chi,\bar{\chi}\right] - \frac{6i\sqrt{a}k}{\kappa}\lambda, \tag{12}$$

$$\bar{S} = \frac{\kappa\pi_a}{\sqrt{a}}\bar{\lambda} + \frac{\kappa\pi_\phi}{\sqrt{a^3}}\bar{\chi} + \frac{\kappa\pi_T}{\sqrt{a^3}}\bar{\eta} - \frac{3i\sqrt{a^3}W}{\kappa}\bar{\lambda} - \frac{i\sqrt{a^3}W_T}{\kappa}\bar{\eta} - \frac{i\sqrt{a^3}W_\phi}{\kappa}\bar{\chi} - \frac{3i\kappa}{4a^{3/2}}\bar{\lambda}\left[\eta,\bar{\eta}\right] - \frac{3i\kappa}{2a^{3/2}}\bar{\lambda}\left[\chi,\bar{\chi}\right] + \frac{6i\sqrt{a}k}{\kappa}\bar{\lambda}. \tag{13}$$

For details on this calculations we refer the reader to [12] where we computed a wave function of the universe for a supersymmetric tachyon in a FRW background, providing a normalizable solution in which the unstable nature of the tachyon can be apreciated; in such study the action (2) was used with the dilaton turned off, whereas in [13] the solution of the Wheeler-DeWitt equation for the complete theory was addressed, where it is introduced a representation for the Grassmann variables in terms of suitable linear combinations of 8-dimension Dirac matrices, therefore the wave function will be an eight-component spinor $\Psi = (\psi_1, \psi_2, \psi_3, \psi_4, \psi_5, \psi_6, \psi_7, \psi_8)$. Upon application of the operators S and $\bar{S}$ on this state, we get two sets of eight equations:

$$\begin{aligned} \left(\hbar\kappa\frac{\partial}{\partial\phi} - \frac{a^3}{\kappa}\frac{\partial W}{\partial\phi}\right)\psi_1 &= 0, \\ \left(\hbar\kappa\frac{\partial}{\partial T} - \frac{a^3}{\kappa}\frac{\partial W}{\partial T}\right)\psi_1 &= 0, \\ \left(\hbar a\kappa^2\frac{\partial}{\partial a} - 3a^3W + 6a^2k + \frac{9}{8}\hbar\right)\psi_1 &= 0, \end{aligned} \tag{14}$$

and

$$\begin{aligned} \left(\hbar\kappa^2\frac{\partial}{\partial\phi} + a^3\frac{\partial W}{\partial\phi}\right)\psi_6 &= 0, \\ \left(\hbar\kappa\frac{\partial}{\partial T} + \frac{a^3}{\kappa}\frac{\partial W}{\partial T}\right)\psi_6 &= 0, \\ \left(\hbar a\kappa^2\frac{\partial}{\partial a} + 3a^3W - 6a^2k + \frac{9}{8}\hbar\right)\psi_6 &= 0, \end{aligned} \tag{15}$$

where we have chosen the particular solution for which $\psi_2 = \psi_3 = \psi_4 = \psi_5 = \psi_7 = \psi_8 = 0$.

Solving these equations we find that

$$\psi_1 = C_1 a^c \exp\left[-\frac{a^{3/2}\left(3\sqrt{a}k - \sqrt{a^3}W(\phi,T)\right)}{\kappa^2\hbar}\right]$$

and

$$\psi_6 = C_2 a^c \exp\left[\frac{a^{3/2}\left(3\sqrt{ak} - \sqrt{a^3}W(\phi,T)\right)}{\kappa^2\hbar}\right],$$

with $c = -\left(\frac{r}{2}+\frac{7}{8}\right)$ being a free parameter arising from the operator ordering. These componentes are normalizable provided that $W(\phi, T \to \pm\infty) > 0$ for ψ_6 and ψ_1 is non-normalizable in this case. On the other hand, if the superpotential $W(\phi, T \to \pm\infty) < 0$ then ψ_6 is non-normalizable and ψ_1 is normalizable, therefore these are non-degenerate eigenstates of the Hamiltonian.

5. Scale factor expectation value

We set $W(\phi,T) = \frac{1}{\sqrt{2}}e^{\phi}\left(\frac{m^2T^4}{24} - \frac{m^2T^3}{12} - \left(\frac{m^4}{160}+\frac{m^2}{240}\right)T^5 + e^T\right)$, this superpotential solution is consistent with the case $k = 0$ for the FRW metric as reported in [5]. Thus we have

$$\psi_6(\phi,T) = a^c \exp\left[\frac{a^3\left(3m^4T^5 + 2m^2T^5 - 20m^2T^4 + 40m^2T^3 - 480e^T\right)}{480\sqrt{2}\kappa^2\hbar}\right],$$

for the normalizable component with $\phi = 0$.

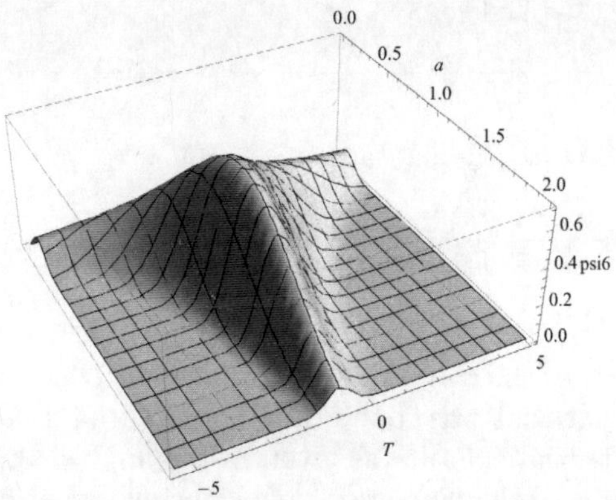

Figure 1. Dependence on a and T of the wave function of the universe.

Regarding T as a time field we can normalize this wave function for $a \in (0,\infty)$, this results in

$$\psi_6(T) = N(T) \exp\left[\frac{a^3\left(3m^4T^5 + 2m^2T^5 - 20m^2T^4 + 40m^2T^3 - 480e^T\right)}{480\sqrt{2}\kappa^2\hbar}\right] \tag{16}$$

where

$$N(T) = \frac{2^{\frac{1}{4}(-3)(2c+1)}3^{\frac{1}{3}-\frac{c}{3}}5^{-\frac{c}{3}-\frac{1}{6}}a^c\left(\frac{-3m^4T^5-2m^2T^5+20m^2T^4-40m^2T^3+480e^T}{\kappa^2\hbar}\right)^{\frac{1}{6}(2c+1)}}{\sqrt{\Gamma\left(\frac{2c}{3}+\frac{1}{3}\right)}}$$

With this wave function we can find the expectation value of the scale factor, $\langle a\rangle$, and the quantum fluctuations, σ_a, explicitly we have

$$\langle a\rangle = \frac{2\sqrt{2}\sqrt[3]{15}\Gamma\left(\frac{2(c+1)}{3}\right)}{\Gamma\left(\frac{2c}{3}+\frac{1}{3}\right)\sqrt[3]{\frac{480e^T - m^2T^3((3m^2+2)T^2-20T+40)}{\kappa^2\hbar}}} \tag{17}$$

and

$$\sigma_a = 2\sqrt[3]{15}\sqrt{\frac{2\Gamma\left(\frac{2c}{3}+\frac{1}{3}\right)\Gamma\left(\frac{2c}{3}+1\right)-2\Gamma\left(\frac{2(c+1)}{3}\right)^2}{\Gamma\left(\frac{2c}{3}+\frac{1}{3}\right)^2\left(\frac{480e^T-m^2T^3(T((3m^2+2)T-20)+40)}{\kappa^2\hbar}\right)^{2/3}}} \tag{18}$$

The behavior of $\langle a\rangle+\sigma_a$, $\langle a\rangle$ and $\langle a\rangle-\sigma_a$ can be seen in the figure below, where the collapse of the universe can be seen directly, which is consistent with the nature of this tachyon potential as stated in [10].

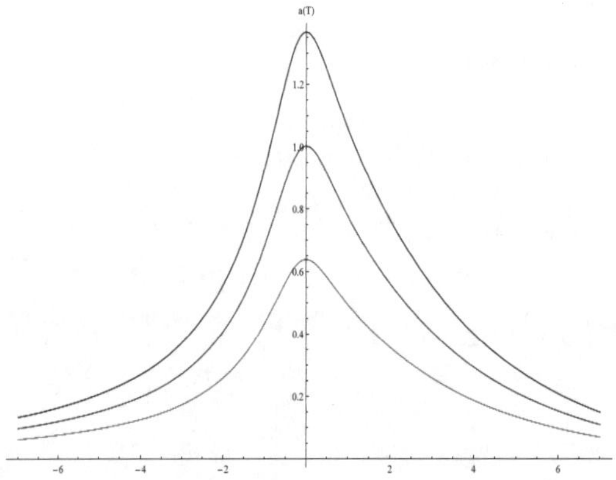

Figure 2. Profile of $\langle a\rangle$ plus fluctuations for $\kappa=1$, $\hbar=1$, $m=1$, $c=1$ and $\phi=0$.

6. Conclusions

We have been capable of extracting classical information from a quantum cosmological model from first principles, by meas of the computation of the expectation value of the scale factor for a FRW model, . The behavior that was found, for the potential used in this work, is consistent with the properties this potential exhibits as given in the analysis by Zwiebach *et al.*, in which the final stage of the evolution of the Universe is a collapsing scenario as the tachyon rolls down. Although we have set $\phi=0$ our study is complete since the roll played by the dilaton is, as expected, that of an overall scale factor. We are working towards obtaining a more detailed analysis of our results and some further predictions. Also, it would be interesting to probe some other proposals for the tachyon potential, specially those with suitable supersymmetric properties.

Acknowledgments

We thank VIEP-BUAP and PIFI-SEP for the support, V. Vázquez-Báez also thanks CONACyT for the studies grant during this work and the organizers of the conference for supporting his attendance.

References

[1] Sen A 2005 *Int. J. Mod. Phys.* A**20** 5513
Sen A 2005 *Phys. Scripta* **T117** 70-75 (*Preprint* hep-th/0312153)
Sen A 2003 *Int. J. Mod. Phys.* A **18**, 4869
Sen A 2002 *Mod. Phys. Lett.* A **17**, 1797 (2002);
Sen A 2002 *J. High Energy Phys.* JHEP**0207**065;
Sen A 2002 *J. High Energy Phys.* JHEP**0204**048
Sen A 2000 *J. High Energy Phys.* JHEP**003**002

IOP Conf. Series: Journal of Physics: Conf. Series **1010** (2018) 012008 doi:10.1088/1742-6596/1010/1/012008

[2] Vafa C 2001 Mirror symmetry and closed string tachyon condensation *Preprint* hep-th/0111051
[3] D'Eath P D 1996 *Supersymmetric quantum cosmology* (Cambridge,UK : Cambridge University Press)
[4] Moniz P V 2010 *Quantum Cosmology - The Supersymmetric Perspective* vol 2 Advanced Topis Lect. Notes Phys. 804 (Berlin: Springer)
[5] García-Jiménez G, Ramírez C and Vázquez-Báez V 2014 *Phys. Rev.* D **89** 043501
[6] Obregón O, Rosales J J, and Tkach V I 1996 *Phys. Rev.* D **53** R1750
[7] Tkach V I, Obregón O, Rosales J J 1997 *Class. Quant. Grav.* **14** 339
[8] Wess J and Bagger J 1992 *Supersymmetry and Supergravity* (Princeton, New Jersey: Princeton University Press)
[9] C. Ramírez 1985 *Ann. Phys.* **186** 43
[10] Yang H and Zwiebach B 2005 *J. High Energy Phys.* JHEP**0508**046
[11] Salam A and Strathdee J 1974 *Nucl. Phys.* **B76**
[12] Vázquez-Báez V and Ramírez C 2013 *AIP Conf. Proc.* **1548** 227 (2013) 9th Mexican School on Gravitational Physics: Cosmology for the 21st Century 3-7 December 2012 (Puerto Vallarta: Jalisco México)
[13] V Vázquez-Báez and C Ramírez 2014 *J.Phys.Conf.Ser.* . **545** 012016 (2014) 10th Workshop of the Gravitation and Mathematical Physics Division of the Mexican Physical Society 2-6 Dec 2013. (Pachuca: Hidalgo México)

IOP Conf. Series: Journal of Physics: Conf. Series **1010** (2018) 012009 doi:10.1088/1742-6596/1010/1/012009

Generalized Entanglement Entropy and Holography

O. Obregón[1]

[1] Departamento de Física, DCI-Campus, León. Universidad de Guanajuato
Loma del Bosque 103, Fracc. Lomas del Campestre, C.P. 37150, Guanajuato, México.

E-mail: `octavio@fisica.ugto.mx`

Abstract.
A nonextensive statistical mechanics entropy that depends only on the probability distribution is proposed in the framework of superstatistics. It is based on a $\Gamma(\chi^2)$ distribution that depends on β and also on p_l. The corresponding modified von Neumann entropy is constructed; it is shown that it can also be obtained from a generalized Replica trick. We address the question whether the generalized entanglement entropy can play a role in the gauge/gravity duality. We pay attention to $2dCFT$ and their gravity duals. The correction terms to the von Neumann entropy result more relevant than the usual UV (for $c = 1$) ones and also than those due to the area dependent AdS_3 entropy which result comparable to the UV ones. Then the correction terms due to the new entropy would modify the Ryu-Takayanagi identification between the CFT entanglement entropy and the AdS entropy in a different manner than the UV ones or than the corrections to the AdS_3 area dependent entropy.

1. Introduction

Information entropies to quantify predictability in a process have been proposed in the literature [1–5]. The most notable of them is due to the seminal work of Shannon [3]. By maximizing the appropriate information measures [1] the associated probability distributions can be calculated. For some of these generalized information and entropy measures their potential physical applications have been discussed elsewhere [6].

By considering nonequilibrium systems with a long-term stationary state that possesses a spatio-temporal fluctuating intensive quantity more general statistics have been formulated known as superstatistics [7]. Selecting the temperature as the fluctuating quantity among other available intensive variables, the microscopic system is considered as made up of smaller cells that are temporarily in local equilibrium. Within each cell β, the inverse temperature, is approximately constant. Each cell is large enough to obey standard statistical mechanics but has a different temperature assigned to it, in accordance to a general distribution $f(\beta)$, from this distribution one can get an effective Boltzmann factor

$$B(E) = \int_0^\infty d\beta f(\beta) e^{-\beta E}, \tag{1}$$

where E is the energy of a microstate associated with each of the considered cells. A distribution $f(\beta) = \delta(\beta - \beta_0)$ gives the ordinary Boltzmann factor. Other distributions, the $\Gamma(\chi^2)$, log-normal and F-distribution have been considered leading to their associated Boltzmann factors.

To deduce the entropies corresponding to these Boltzmann factors $B(E)$ a formalism was introduced [8]. Following this procedure the Boltzmann-Gibbs entropy and the non-extensive statistical mechanics entropy were obtained. The entropy is expressed as $S = k\sum_{k=1}^{\Omega} s(p_l)$ in terms of a generic $s(x)$ that can be calculated, x is at this stage a parameter. By maximizing the appropriate information measure this parameter is identified with the probability distribution. For other distributions, among them, the log-normal and F-distribution it is not possible to get closed analytical expressions for their associated entropies and the calculations were performed numerically [8] by means of the corresponding $B(E)$.

In [9] a $f(\beta)$, $\Gamma(\chi^2)$ distribution was proposed depending on a parameter p_l, from it the associated Boltzmann factor $B(E)$ and entropy were calculated. The parameter p_l is then identified with the probability distribution by maximizing the appropriate information measure. The resulting generalized information entropy depends only on the probability distribution. This new entropy was further studied in [10]. In this paper, we will review some of our previous results in [9, 10]. We will also present some new results which correspond to work in progress with several collaborators and that will be further developed and presented elsewhere. First, in Section 2 several generalized, p_l dependent distributions and their associated Boltzmann factors are presented. It is shown that they coincide up to the first correction term in their expansions. The one corresponding to the $\Gamma(\chi^2)$ distribution can be obtained in an analytically closed form. In Section 3 we will obtain the entropy for the $\Gamma(\chi^2)$ distribution. We will show several aspects of this non-extensive generalized entropy that depends only on the probability.

We turn to quantum aspects in Section 4, and construct the corresponding generalization of the von Neumann entropy and exhibit an extended Replica trick from which this same entropy also arises. We address then the question whether an alternative entanglement entropy can play a role in the gauge/gravity duality. In particular, based on our generalized entropy we obtain correction terms to the usual $2dCFT$ entanglement entropy. These are exponentially suppressed, but are larger than the standard ultraviolet terms, that are also exponentially suppressed. We point out to the possible implications of this entropy in connection to the Ryu-Takayanagi identification between the CFT entanglement entropy and their proposed AdS_3 area dependent entropy.

2. Generalized distributions and their Boltzmann factors

We begin by assuming a Γ (or χ^2) distribution on the inverse temperature β, depending on a parameter p_l to be identified with the probability associated with the microscopic configuration of the system by means of maximizing the associated entropy. We may write this parameter p_l Γ distribution as

$$f_{p_l}(\beta) = \frac{1}{\beta_0 p_l \Gamma\left(\frac{1}{p_l}\right)}\left(\frac{\beta}{\beta_0}\frac{1}{p_l}\right)^{\frac{1-p_l}{p_l}} e^{-\beta/\beta_0 p_l}, \tag{2}$$

where β_0 is the average inverse temperature. Integration over β yields the generalized Boltzmann factor

$$B_{p_l}(E) = (1 + p_l\beta_0 E)^{-\frac{1}{p_l}}, \tag{3}$$

as shown in [10], this kind of expression can be expanded for small $p_l\beta_0 E$, to get

$$B_{p_l}(E) = e^{-\beta_0 E}\left[1 + \frac{1}{2}p_l\beta_0^2E^2 - \frac{1}{3}p_l^2\beta_0^3E^3 + ...\right]. \tag{4}$$

We follow now the same procedure for the log-normal distribution, this can be written in terms of p_l as

$$f_{p_l}(\beta) = \frac{1}{\sqrt{2\pi}\beta[\ln(p_l+1)]^{1/2}} \exp\left\{ -\frac{\left[\ln \frac{\beta(p_l+1)^{1/2}}{\beta_0}\right]^2}{2\ln(p_l+1)} \right\}. \tag{5}$$

The generalized Boltzmann factor can be obtained in leading order, for small variance of the inverse temperature fluctuations,

$$B_{p_l}(E) = e^{-\beta_0 E}\left[1 + \frac{1}{2}p_l\beta_0^2E^2 - \frac{1}{6}p_l^2(p_l+3)\beta_0^3E^3 + \cdots\right]. \tag{6}$$

In general, the F-distribution has two free constant parameters. We consider, the particular case in which one of these constant parameters is chosen as $v = 4$. For this value of this constant parameter we define a F-distribution in function of the inverse of the temperature and p_l as

$$f_{p_l}(\beta) = \frac{\Gamma\left(\frac{8p_l-1}{2p_l-1}\right)}{\Gamma\left((\frac{4p_l+1}{2p_l-1}\right)}\frac{1}{\beta_0^2}\left(\frac{2p_l-1}{p_l+1}\right)^2 \frac{\beta}{\left(1+\frac{\beta}{\beta_0}\frac{2p_l-1}{p_l+1}\right)^{\left(\frac{8p_l-1}{2p_l-1}\right)}}. \tag{7}$$

Once more the associated Boltzmann factor can not be evaluated in a closed form, but for a small variation of the fluctuations we obtain the series expansion

$$B_{p_l}(E) = e^{-\beta_0 E}\left[1 + \frac{1}{2}p_l\beta_0^2E^2 + \frac{1}{3}p_l\frac{(5p_l-1)}{p_l-2}\beta_0^3E^3 + \ldots\right]. \tag{8}$$

As we will demonstrate and already shown in [9, 10] one can obtain in a closed form the entropy corresponding to Eqs. (2) and (3) resulting in

$$S = k\sum_{l=1}^{\Omega}(1 - p_l^{p_l}), \tag{9}$$

where k is the conventional constant and $\sum_{l=1}^{\Omega} p_l = 1$. The expansion of Eq. (9) gives

$$-\frac{S}{k} = \sum_{l=1}^{\Omega} p_l \ln p_l + \frac{(p_l \ln p_l)^2}{2!} + \frac{(p_l \ln p_l)^3}{3!} + \cdots. \tag{10}$$

Given that the Boltzmann factor coincides up to the second term for the Γ (χ^2), log-normal and F-distributions, for enough small $p_l\beta_0 E$ the entropy Eq. (10) corresponds to all these distributions up to the first term that modifies the usual Shannon entropy. We expect at least this modification to the entropy for several possible $f(\beta)$ distributions. So, for an stationary system with some temperature distribution in the mentioned sense we could describe the actual β -distributions by some of the ones considered here. As up to the first correction term to the standard Boltzmann factor is the same for several distributions, as shown, also the first correction term to the various corresponding entropies will be the same, namely the second term in Eq.(10).

3. Entropy from the Boltzmann factor

We begin by assuming the Γ (or χ^2) inverse temperature β distribution depending on a parameter p_l Eq. (2), to be identified with the probability associated with the microscopic configuration of the system. As shown, integration over β yields the generalized Boltzmann factor which can be expanded as in Eq. (4).

The examples studied in [7] have been nicely addressed in [8] in order to deduce the entropies from their corresponding Boltzmann factors. Another possible way to reconstruct entropies

IOP Conf. Series: Journal of Physics: Conf. Series **1010** (2018) 012009 doi:10.1088/1742-6596/1010/1/012009

depending on constant parameters has been proposed in [11–13], this approach provides other expressions. In [12,13] it has been shown that there exists a duality between these two procedures. We will restrict to the first proposal, by means of which the Boltzmann-Gibbs entropy and the non extensive statistical mechanics entropy can be obtained in a closed analytic form [8]. However, the entropies corresponding to the Boltzmann factors associated to the log-normal or the F-distributions can not be obtained analytically and were calculated numerically. Following [7,8], we present the procedure to obtain the entropy corresponding to the $f(\beta)$ distribution Eq. (2) and to its associated generalized Boltzmann factor Eq. (3). We begin by defining the entropy $S = k\sum_{l=1}^{\Omega} s(p_l)$ in terms of a generic $s(p_l)$, where p_l can be considered at this stage an arbitrary parameter; for $f(\beta) = \delta(\beta - \beta_0)$ $s(x) = -x\ln x$ the Shannon entropy is recovered. As shown in [8] it is possible to express $s(x)$ and a generic internal energy $u(x)$ in terms of integrals on a function $E(y)$ that is obtained from the Boltzmann factor $B(E)$ of interest. By these means $s(x)$ and $u(x)$ can be written as

$$s(x) = \int_0^x dy \frac{\delta + E(y)}{1 - E(y)/E^*}\,, \tag{11}$$

and

$$u(x) = (1 + \delta/E^*)\int_0^x \frac{dy}{1 - E(y)/E^*}\,, \tag{12}$$

where $E(y)$ is to be identified with the inverse function of $B_{p_l}(E)/\int_0^\infty dE'\, B_{p_l}(E')$. One selects first the $f(\beta)$ of interest, then $B(E)$ is calculated and the integral $\int_0^\infty B(E')\,dE'$ is performed. Inverting the axes of the variables, $E(y)$ for several superstatistics can be found [8], and from it E^*. In our case, the starting points are the distribution Eq. (2) and the Boltzmann factor Eq. (3). $E(y)$ and E^* are given by

$$E(y) = \frac{y^{-x} - 1}{x}\,, \tag{13}$$

$$E^* = -\frac{1}{x}. \tag{14}$$

A straightforward calculation gives for $u(x)$ and $s(x)$

$$u(x) = x^{x+1}, \tag{15}$$

$$s(x) = 1 - x^x. \tag{16}$$

where δ has been determined by means of the condition $u(1) = 1$.

Expressions Eqs.(15) and (16) fulfill the expected conditions for the entropy and the energy $s(0) = 0, u(0) = 0$ and $u(1) = 1, s(1) = 0$. By these means, the entropy results in Eq.(9) and its expansion in Eq.(10).

Using these results, the corresponding functional including restrictions is given by

$$\Phi = \frac{S}{k} - \gamma\sum_{l=1}^{\Omega} p_l - \beta\sum_{l=1}^{\Omega} p_l^{p_l+1} E_l, \tag{17}$$

where the second restriction concerns the average value of the energy and γ and β are Lagrange parameters, and then by maximizing Φ, p_l is obtained as

$$1 + \ln p_l + \beta E_l(1 + p_l + p_l \ln p_l) = p_l^{-p_l}. \tag{18}$$

The dominant term in this expression corresponds to the Gibbs-Boltzmann prediction, $p_l = e^{-\beta_o E_l}$. In general, however, we cannot analytically express p_l as function of βE_l. In Fig. 1, p_l is given as a function of the reduced energy βE_l. We notice that for relative large values of βE_l the usual values for p_l coincide with the ones given by Eq. (18). As expected, they coincide also for $p_l \sim 1$.

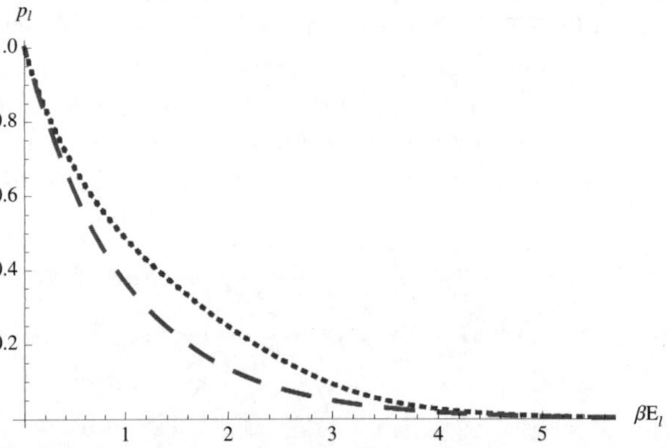

Figure 1. (Color online) Comparison of the two probabilities. The blue dotted line corresponds to $p_l = f(\beta E_l)$, Eq. (18), and red dashed line to the standard one $p_l = e^{-\beta E_l}$

As we have shown, by choosing $f_{p_l}(\beta)$ Eq.(2), then $B_{p_l}(E)$ Eq. (3) is obtained by integrating over β; by inverting the axes of the variable the inverse function $E(y)$ Eq. (13) and $E^\star$ Eq. (14) can be found. This procedure has allowed us to calculate $u(x)$ Eq. (15) and $s(x)$ Eq. (16) and consequently the entropy Eqs.(9) and (16). If we assume in $f_{p_l}(\beta)$ Eq. (2) the equiprobable condition, $p_l = \frac{1}{\Omega}$, then the corresponding distribution is given by

$$f_\Omega(\beta) = \frac{\Omega}{\beta_0 \Gamma(\Omega)} \left(\frac{\beta\Omega}{\beta_0}\right)^{\Omega-1} e^{-\frac{\beta\Omega}{\beta_o}}, \tag{19}$$

where the Boltzmann factor and the entropy are now

$$B_\Omega(E) = (1 + \beta_0 E/\Omega)^{-\Omega}, \tag{20}$$

$$S = k\Omega \left[1 - \frac{1}{\Omega^{\frac{1}{\Omega}}}\right], \tag{21}$$

or, in terms of Boltzmann's entropy, $S_B = k \ln \Omega$,

$$\frac{S}{k} = \frac{S_B}{k} - \frac{1}{2!} e^{-S_B/k} \left(\frac{S_B}{k}\right)^2 + \frac{1}{3!} e^{-\frac{2S_B}{k}} \left(\frac{S_B}{k}\right)^3 \cdots . \tag{22}$$

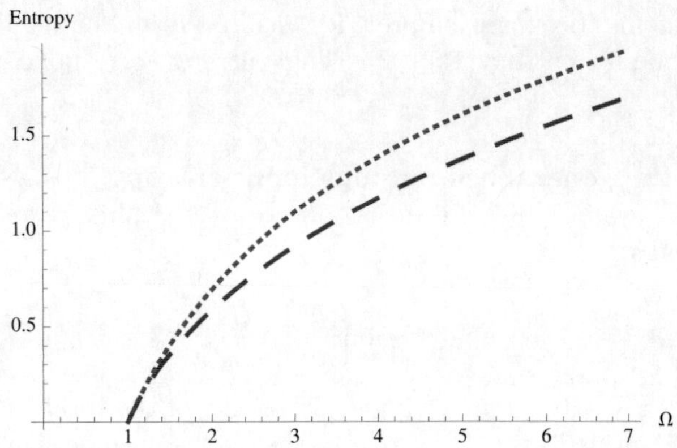

Figure 2. (Color online) The entropies as function of Ω (small)the blue dashed and red dotted lines correspond to $\frac{S}{k}$, Eq.(21) and, $\frac{S_B}{k}$, respectively ($p_l = 1/\Omega$ equipartition)

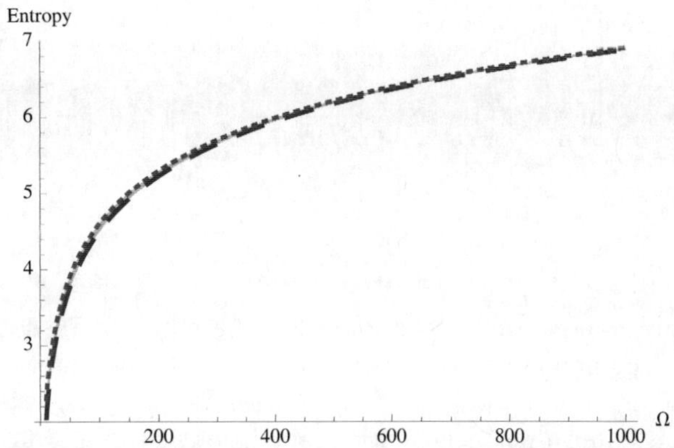

Figure 3. (Color online) The entropies as function of Ω (large) the blue dashed and red dotted lines correspond to $\frac{S}{k}$, Eq. (21) and $\frac{S_B}{k}$, respectively ($p_l = 1/\Omega$ equipartition)

Figs. 2 and 3 show the Boltzmann entropy $\frac{S_B}{k}$ and the entropy $\frac{S}{k}$ given by expression Eq. (21). As mentioned in the Introduction, it was shown in [9], in relation with the entropy of a black hole, that if we associate its entropy, depending linearly on its area, with $\frac{S_B}{k}$ the standard entropy, then the entropy $\frac{S}{k}$ will be given as a function of the area by means of Eqs. (21) and (22). This would imply a modification to Newton's law and to general relativity according with the possibility that gravity could be thought as an equation of state [14]; explained as an entropic force [15–17]. However, for most gravitational systems one expects a large Ω, namely large $\frac{S_B}{k}$ and will not greatly differ with $\frac{S}{k}$, Eq. (21) and Fig. 3.

We notice that in the range of low values of Ω the entropies $\frac{S_B}{k}$ and $\frac{S}{k}$ differ. Instead, for large Ω the two entropies essentially coincide. Since Ω is basically a measure of the phase space volume, what we are finding is that for systems with reduced number of microstates the entropies become different, whereas for the opposite case they will be essentially identical. Then, we could conclude from here that the model could be useful when we have a clear indication of restriction of available states, like by strong confinement of fluids or low temperatures.

The entropy derived in [9,10] and whose main properties we review and extended here, Eqs.(9) and (21), has as an important feature to be independent of any arbitrary constant parameter, and to depend only on the probability distribution p_l Eq.(9), associated with the microscopic configuration of the system. Its expansion provides as a first term Shannon's entropy Eq.(10) and correspondingly Boltzmann's entropy Eq.(22). This entropy corresponds to the Γ distribution Eqs.(2) and (3).

4. Holography and the generalized entanglement entropy

A quantum many body system in one dimension and at criticality corresponding to a $2dCFT$ has an entanglement entropy [18] given by

$$S_A = \frac{c}{3}\ln\left(\frac{L}{\pi a}\sin\frac{\pi l}{L}\right), \tag{23}$$

where l, is the length of the subsystem, A and L the length of the total system AB, both ends of AUB are periodically identified, a is a UV cutoff, the lattice spacing, is the central charge c of the $2dCFT$. Away from criticality the previous expression [18] is replaced by

$$S_A = \frac{1}{3}\ln\frac{l}{a}, \tag{24}$$

where for simplicity we have putted $c = 1$. The UV correction terms to (24) have been calculated [19] and can be written as

$$S = \frac{1}{3}\ln\frac{l}{a} - 2\left(1 + \ln\left(\frac{a}{l}\right)\frac{\partial}{\partial\ln(\frac{a}{l})}\right)\sum_{k=1}^{\infty}\ln\left(1 - \left(\frac{a}{l}\right)^{2k}\right), \tag{25}$$

in terms of S_A and for $\frac{a}{l} \ll 1$,

$$S \sim S_A - 12S_A e^{-6S_A} + 2e^{-6S_A} + \cdots, \tag{26}$$

the correction terms are exponentially suppressed. The entropy S_A corresponds to the most relevant term, the von Neumann entropy for the $2dCFT$ under consideration.

The entropy in AdS_3 is proportional to a minimal surface [20]. In this case the length of an static geodesic determines its value. A geodesic going from radius ρ_0 and $\theta = 0$ to radius ρ_0 and $\theta = \frac{\pi l}{L}$ has a length given by

$$L_{\gamma_A} = Rarccosh\left[1 - sin^2\left(\frac{l\pi}{L}\right) + \frac{1}{2}e^{2\rho_0}sin^2\left(\frac{l\pi}{L}\right) + \frac{1}{2}e^{-2\rho_0}sin^2\left(\frac{l\pi}{L}\right)\right]. \tag{27}$$

For enough large ρ_0 and identifying

$$e^{\rho_0} \equiv \frac{L}{\pi a}, \qquad \frac{R}{4G} \equiv \frac{c}{6}. \tag{28}$$

It was shown that the length Eq.(27) provides an area entropy dependent that coincides with Eq.(24). Further terms can be calculated from the length Eq.(27), these result in the correction terms to the AdS_3 entropy. It can be shown that they are proportional to the same negative exponentials than the $2dCFT$ entropy UV correction terms Eq.(26).

We turn now to the generalization of the von Neumann entropy that corresponds to the proposed generalized Shannon entropy Eq.(9), namely

$$S_+ = T_r(1 - \rho^\rho), \tag{29}$$

in which ρ is the density matrix. This same entropy arises also from the following generalized Replica trick that we propose [21, 22]

$$S_+ = -\sum_{k\geq 1} \frac{1}{k!} \lim_{n\to k} \frac{\partial^k}{\partial n^k} Tr\rho_A^n. \tag{30}$$

Making n-copies of the system is possible to compute $Tr\rho_A^n$ for n a positive integer. The partition function of the system with this n-sheeted structure reads [18] $Z_n(A)$ with

$$Tr\rho_A^n = \frac{Z_n(A)}{Z^n}. \tag{31}$$

For all the examples of $2dCFT$ using the transformation property of the two-point functions when the lattice spacing is very small one has a generic expression

$$Tr\rho_A^n \sim b^{(l/n-n)/6}. \tag{32}$$

We show here the simplest case, for $b = \frac{l}{a}$ and then the standard Replica trick, namely

$$S_A = -\frac{\partial}{\partial n} Tr\rho_A^n|_{n=1} \tag{33}$$

reproduces the entropy S_A Eq.(24).

Further terms in the entropy Eq.(29), are given by the second and third derivatives in Eq.(30) and the S_+ entropy results in

$$S_+ = S_A + \frac{1}{16} e^{-\frac{3}{4}S_0} S_0 \left(1 + \frac{25}{8} S_0\right) - \frac{1}{6} S_0 e^{-4S_0/3} \left(\frac{1}{27} + \frac{5}{181} S_0 + \frac{125}{729} S_0^3\right) + \cdots \tag{34}$$

We notice that the correction terms to S_A due to the proposed generalized entanglement entropy Eq.(29) are also exponentially suppressed. They are however larger than the usual ultraviolet corrections to the $2dCFT$ entanglement entropy Eq.(26) and, as mentioned, consequently also larger than the correction entropy terms to the AdS_3 space. The proposed entanglement entropy Eq.(29) has its origin in other considerations Eqs.(9) and (30). It seems of interest to search for an entropy for the AdS_3 space that would correspond to the entropy S_+ Eq.(29). This is not a straightforward task one would need to find a well justified and appropriate generalization of the AdS_3 space entropy based on a possibly different relation with the geodesical length Eq.(27) and probably then another theory of gravitation would be needed. These matters require further study and are beyond the scope of this work.

Acknowledgements
Thanks are expressed to Nana Cabo, Jose Torres and Alejandro Gil, my collaborates. Further results will be published elsewhere. I also acknowledge useful comments by Ramón Castañeda, Marco Ortega and Salomón Zacarías. This work was supported by CONACYT project 257919 and CA40130116, PRODEP and UG projects.

References

[1] A. Rényi, *Probability theory*, North Holland, Amsterdam (1970); G. Kaniadakis, Phys. Rev. E 66, 056125 (2002); C. Tsallis, J. Stat. Phys. 52, 479 (1988); S. Abe, Phys. Lett. A 224, 326 (1997); B.D. Sharman, D.P. Mittal, J. Math. Sci. 10, 28 (1975); M.D. Esteban, D. Morales, Kybernetica 31, 337 (1995); A.N. Gorban, P.A. Gorban, G. Judge, Entropy 12 (5), 1145 (2010).
[2] A.M. Crawford, N. Mordant, and E. Bodenschatz, *Dynamical Foundations of Nonextensive Statistical Mechanics.* [physics/0212080].
[3] C.E. Shannon, Bell. Syst.Tech. J. 27, 623 (1948).
[4] S. Abe, C. Beck and E.G.D. Cohen, Phys. Rev. E 76, 031102 (2007); P.H. Chavanis, Eur. Phys. J.B 62, 179 (2008).
[5] C. Beck, *Generalized information end entropy measure in physics.* [arXiv:0902.1235].

IOP Conf. Series: Journal of Physics: Conf. Series **1010** (2018) 012009 doi:10.1088/1742-6596/1010/1/012009

[6] G. Wilk, Z. Wlodarczyk, Phys. Rev. Lett. 84, 2770 (2000); H. Sakaguchi, J. Phys. Soc. Japan 70, 3247 (2001); S. Jung, H.L. Swinney, Superstatistics in Taylor Couette Flow, University of Austin (2002) [prepint].
[7] C. Beck, E.G.D. Cohen, Physica A 322, 267 (2003).
[8] C. Tsallis, A.M.C. Souza, Phys. Rev. E 67, 026106 (2003).
[9] O. Obregón, Entropy 12 (9), 2067 (2010).
[10] O. Obregón and A. Gil-Villegas, Phys. Rev. E 88, 062146 (2013).
[11] R. Hanel and S. Thurner, Physica A 380, 109 (2007).
[12] R. Hanel and S. Thurner, *Entropies for complex systems: Generalized-generalized entropies, Complexity, Metastability and Nonextensivity*, (Am. Inst. of Physics, College Park, MD, 2007), p. 68-75.
[13] R. Hanel, S. Thurner and M. Gell-Mann, PNAS 108 (2011) 6390; R. Hanel, S. Thurner and M. Gell-Mann, PNAS 109, 19151 (2012).
[14] T. Jacobson, Phys. Rev. Lett. 75, 1260 (1995).
[15] E.P. Verlinde, JHEP 1104, 029 (2011). [arXiv:1001.0785]
[16] J.C. López Domínguez, O. Obregón, M. Sabido and C. Ramírez, Phys. Rev. D 74, 084024 (2006).
[17] A. Sheykhi, Phys. Rev D 81, 104011 (2010).
[18] Calebrese, P. and Cardy, J., *Entanglement entropu and quantum field theory*, Journal of Statistical Mechanics: Theory and Experiment 2004, no. 06, 06002 (20014).
[19] Holzhey, F, Larsen, F and Wilczek, F. Nucl. Phys. B 424, 443 (1994).
[20] Ryu, S. and Takayanagi, T., Phys. Rev. Lett. 96, 181602 (2006).
[21] Cabo, N. and O. Obregón, *Generalized entanglement entropy and holography*, arXiv:1507.00779.
[22] O. Obregón, Int.J.Mod.Phys. A30 (2015) No.16, 1530039.

IOP Conf. Series: Journal of Physics: Conf. Series **1010** (2018) 012010 doi:10.1088/1742-6596/1010/1/012010

Group theoretical approach to the Dirac operator on S^2

Sergio Gutiérrez[1] **and Idrish Huet**[2]

[1] Departamento de Física, Universidad Autónoma Metropolitana–Iztapalapa, Apartado Postal 55–534, C.P. 09340, México, D.F.

[2] Facultad de Ciencias en Física y Matemáticas, Universidad Autónoma de Chiapas, Ciudad Universitaria, Tuxtla Gutiérrez 29050, México

E-mail: sergiogs@xanum.uam.mx, idrish@ifm.umich.mx

Abstract. In this revision we outline the group theoretical approach to formulate and solve the eigenvalue problem of the Dirac operator on the round 2-sphere conceived as the right coset $S^2 = SU(2)/U(1)$. Starting from general symmetry considerations we illustrate the formulation of the Dirac operator through left action or right action differential operators, whose properties on a right coset are quite different. The construction of the spinor space and the solution of the spectral problem using group theoretical methods is also presented.

1. Introduction

Besides their continued use in fundamental quantum field theories, Dirac operators on curved spaces have sparked renewed interest due to other reasons, notably their use in condensed matter physics where, amongst other applications, they are central to the effective field description of graphene, fullerenes and the quantum Hall effect [1, 2]. These are both examples of Dirac operators on a Riemannian manifold, wherein no timelike coordinate exists on account of the metric signature, and where the square of the Dirac operator may be interpreted as the hamiltonian of a non-relativistic fermion confined to a curved space, out of which S^2 stands out as one of the simplest examples. There are presently several different accounts of the Dirac operator on S^2 given in the literature, most remarkably [3–9], many of which rely on the standard and general approach of vielbeins. However, despite the generality of the vielbein method it is known that when we deal with a coset space G/H an elegant and natural treatment that takes advantage of the symmetry is available cf. [10]; this is the so called group theoretical approach and relies mainly on representation theory. In what follows we shall review the essential aspects of the group theoretical approach to the Dirac operator's spectral problem on $S^2 = SU(2)/U(1)$.

It has been established [11] that for a simply connected spin manifold $\mathcal{M}$ with riemannian metric tensor $g_{\mu\nu}$ and vielbeins e^i_μ, such as S^2, the Dirac operator corresponding to the (unique) spin structure is

$$\not{D} = i\gamma^\mu \nabla_\mu \tag{1}$$

where $\nabla_\mu = \partial_\mu + \Omega_\mu$ is the covariant spinor derivative and $\Omega_\mu = -\frac{1}{8} e^i{}_\sigma \hat{\nabla}_\mu e^{j\sigma} [\gamma_i, \gamma_j]$ is the spin connection corresponding to the spin structure of $\mathcal{M}$. The euclidean Dirac matrices introduced satisfy $\{\gamma_i, \gamma_j\} = 2\delta_{ij}\mathbf{1}$, while the riemannian Dirac matrices $\gamma_\mu = e^j_\mu \gamma_j$ accordingly

$$\{\gamma^\mu, \gamma^\nu\} = 2g^{\mu\nu}\mathbf{1}. \tag{2}$$

As mentioned before, when dealing with a coset space $\mathcal{M} = G/H$ the Dirac operator and its eigenvalue problem take a particularly simple form in terms of representation theory. In general given a finite dimensional Lie group G and a Lie subgroup $H \subset G$, with Lie algebras $\mathfrak{g}$ and $\mathfrak{h} \subset \mathfrak{g}$, one can decide on constructing either a left coset space $G\backslash H$ or a right coset space G/H, being both equivalent in a sense; but once a choice has been made then the left and right group actions will be radically distinct. Essentially on a right coset $G/H = \{[g] = gH : g \in G\}$ the left infinitesimal actions of G give rise to Lie derivatives and the right actions to covariant derivatives. Let T_a be the generators of $\mathfrak{g}$, we shall assume from now on that G is compact, and hence isomorphic to a group of unitary matrices, therefore we loose no generality considering that $T_a^\dagger = T_a$ as complex matrices, and setting the Lie algebra to be[1]

$$[T_a, T_b] = i f_{abc} T_c \tag{3}$$

with real structure constants f_{abc}. Given a function $f \in L^2(G, \mathbb{C})$ when G is compact the Peter-Weyl theorem [12] asserts that it can be expanded in the orthogonal basis formed by the elements of the Wigner D-matrices corresponding to all the unitary inequivalent irreducible representations D^J of G (i.e. the representation ring $\widehat{G}$, labelled by J).

$$f(g) = \sum_{J\in\widehat{G}} \sum_{i,k\in I(J)} f^J_{ik} D^J_{ik}(g), \quad I(J) \text{ index set of } D^J, \quad f^J_{ik} \in \mathbb{C} \tag{4}$$

on the other hand, a function on the coset space $\varphi : G/H \to \mathbb{C}$ can be naturally understood as a function over G that takes constant values on each coset class, such property is called H-equivariance, and is tantamount to $\varphi(gh) = \varphi(g)$ for each $h \in H$, it follows then that such functions must be expanded in the form [5]

$$\varphi(g) = \sum_{L\in\widehat{G/H}} \sum_{\substack{i\in I(L)\\ k_0\in L_0}} \varphi^L_{ik_0} D^L_{ik_0}(g), \qquad \varphi^L_{ik_0} \in \mathbb{C}. \tag{5}$$

Above $\widehat{G/H} \subset \widehat{G}$ denotes the subset of irreducible representations that, when reduced under H, contain the trivial representation at least once, while the subset $L_0 \subset I(L)$ labels such trivial subrepresentation of H within D^L. Typical examples of the expansion (5) include Fourier series, expansions in spherical harmonics, and harmonic expansions in general. Left and right actions can be defined on a group $L, R : G \times G \to G,\ (k, g) \mapsto L_k g, R_k g$, namely

$$L_k g := kg, \quad R_k g := gk^{-1}\ . \tag{6}$$

Such maps are compatible with G in that $R_g \circ R_k = R_{gk}$, $L_g \circ L_k = L_{gk}$. Let a differentiable function be given $f : G \to \mathbb{C}$, define now the left action differential operators $\mathcal{L}_a$ and the right action differential operators $\mathcal{R}_a$ as the vector fields

$$i\mathcal{L}_a f(g) := \left.\frac{df(exp(-isT_a)g)}{ds}\right|_{s=0}, \quad i\mathcal{R}_a f(g) := \left.\frac{df(gexp(+isT_a))}{ds}\right|_{s=0} \tag{7}$$

where exp is the exponential map as usual. These differential operators form a representation of $\mathfrak{g}$ with the same commutation relations

[1] Since G is compact we have written f_{abc}, as the structure constants are totally antisymmetric tensors and no distinction is required between upper and lower indices.

$$[\mathcal{L}_a, \mathcal{L}_b] = i f_{abc} \mathcal{L}_c, \quad [\mathcal{R}_a, \mathcal{R}_b] = i f_{abc} \mathcal{R}_c. \tag{8}$$

We would like to point out that although functions on the coset $\varphi : G/H \to \mathbb{C}$ are just a subclass of functions on G, namely the H-equivariant functions, the left action is well defined on such functions φ, but the right action is not well defined since its effect depends on the representative chosen within each coset class. In an explicit matrix representation, cf. [13], it has been found that the left and right action operators are

$$\mathcal{L}_a = -\mathrm{Tr}\left(T_a g \frac{\partial}{\partial g^{\mathrm{T}}} \right), \quad \mathcal{R}_a = \mathrm{Tr}\left(g T_a \frac{\partial}{\partial g^{\mathrm{T}}} \right), \tag{9}$$

notice that they are dual to the left and right Maurer-Cartan forms $\theta_L, \theta_R \in \Omega^1(G) \otimes \mathfrak{g}$:

$$\theta_L = -(dg)g^{-1}, \quad \theta_R = g^{-1}dg, \quad \langle \theta_L, \mathcal{L}_a \rangle = \langle \theta_R, \mathcal{R}_a \rangle = T_a \tag{10}$$

2. The round sphere S^2

It is clear that the Dirac operator depends on the metric geometry of the space, we shall consider a "round" metric on the sphere, i.e. the $SU(2)$-invariant metric induced by the G-invariant metric [14] on $SU(2)$. To this end a parametrisation for $SU(2)$ is chosen with global coordinates $z_A, \bar{z}_B \in \mathbb{C}$, $A, B = 1, 2$ subjected to a constraint

$$g = \begin{pmatrix} z_1 & -\bar{z}_2 \\ z_2 & \bar{z}_1 \end{pmatrix}, \quad |z_1|^2 + |z_2|^2 = 1 \tag{11}$$

the round sphere can be seen embedded in $\mathbb{R}^3$ through the cartesian coordinates $(x_1, x_2, x_3) = x$ which relate to g through the projection of the Hopf bundle $\pi : S^3 \to S^2$, $z \mapsto x$ given by

$$x_a = z^\dagger \sigma_a z, \quad z = \begin{pmatrix} z_1 \\ z_2 \end{pmatrix}, \quad x \cdot x = 1. \tag{12}$$

The constraint on x follows directly from the constraint on z, that is $z^\dagger z = 1$, and σ_a are the Pauli matrices. The embedding $U(1) \hookrightarrow SU(2)$ we shall use is the diagonal embeding, so that $U(1)$ is generated by σ_3

$$h = \begin{pmatrix} e^{i\alpha} & 0 \\ 0 & e^{-i\alpha} \end{pmatrix} \in U(1) \subset SU(2) \tag{13}$$

Henceforth we shall denote by D^J the spin J irreducible representation of $SU(2)$, in particular the expansion (5) becomes the familiar spherical harmonic expansion on the sphere

$$\varphi(g) = \sum_{L=0}^{\infty} \sum_{M=-L}^{L} \varphi^L_{M0} D^L_{M0}(g), \quad D^L_{M0}(g) = Y_{LM}(x) \tag{14}$$

where the D-Wigner matrices $D^J_{MN}(g)$ may be found e.g. in [15], in this case we have set $i = M$ and recognised $\widehat{S^2} = \mathbb{N} \subset \widehat{SU(2)} = \frac{1}{2}\mathbb{N}$, $L_0 = \{0\}$ and $I(L) = \{-L, \cdots, 0, \cdots, L\}$. Consequently the representation content of smooth functions is clearly

$$C^\infty(S^2, \mathbb{C}) \subset \bigoplus_{J=0}^{\infty} D^J\,. \tag{15}$$

The Maurer-Cartan forms used in appendix Appendix A to find the Ricci curvature are

$$\theta_L = -\begin{pmatrix} \bar{z}_1 dz_1 + z_2 d\bar{z}_2 & \bar{z}_2 dz_1 - z_1 d\bar{z}_2 \\ \bar{z}_1 dz_2 - z_2 d\bar{z}_1 & \bar{z}_2 dz_2 + z_1 d\bar{z}_1 \end{pmatrix}, \; \theta_R = \begin{pmatrix} \bar{z}_1 dz_1 + \bar{z}_2 dz_2 & \bar{z}_2 d\bar{z}_1 - \bar{z}_1 d\bar{z}_2 \\ z_1 dz_2 - z_2 dz_1 & z_2 d\bar{z}_2 + z_1 d\bar{z}_1 \end{pmatrix}. \tag{16}$$

The G-invariant metric on G, and the corresponding line element are related to the Maurer-Cartan forms, taking (11) into account one finds

$$ds^2 = -\frac{1}{2}\mathrm{Tr}[(\theta_L)^2] = -\frac{1}{2}\mathrm{Tr}[(\theta_R)^2] = dz_1 d\bar{z}_1 + dz_2 d\bar{z}_2. \tag{17}$$

We may now write the Dirac operator using either left or right action differential operators, introducing the complex basis $\mathcal{R}_\pm = \mathcal{R}_1 \pm i\mathcal{R}_2$, $\sigma_\pm = \frac{1}{2}(\sigma_1 \pm i\sigma_2)$, $\mathcal{R}_0 = \mathcal{R}_3$, $\sigma_0 = \sigma_3$

$$\mathcal{L}_+ = \bar{z}_1\bar{\partial}_2 - z_2\partial_1, \quad \mathcal{L}_- = \bar{z}_2\bar{\partial}_1 - z_1\partial_2, \quad \mathcal{L}_0 = \frac{1}{2}(z_2\partial_2 - z_1\partial_1 + \bar{z}_1\bar{\partial}_1 - \bar{z}_2\bar{\partial}_2), \tag{18}$$

$$\mathcal{R}_+ = z_2\bar{\partial}_1 - z_1\bar{\partial}_2, \quad \mathcal{R}_- = \bar{z}_1\partial_2 - \bar{z}_2\partial_1, \quad \mathcal{R}_0 = \frac{1}{2}(z_1\partial_1 + z_2\partial_2 - \bar{z}_1\bar{\partial}_1 - \bar{z}_2\bar{\partial}_2), \tag{19}$$

in the following manner, cf. [4, 5] :

$$\not{D}_L = \sigma_a\mathcal{L}_a + 1, \quad \not{D}_R = \sigma_+\mathcal{R}_- + \sigma_-\mathcal{R}_+. \tag{20}$$

It should be noted that while $\mathcal{L}_a : C^\infty(S^2, \mathbb{C}) \to C^\infty(S^2, \mathbb{C})$, the right operators $\mathcal{R}_\pm$ map smooth functions to (anti)holomorphic vector fields on the sphere and $\mathcal{R}_0$ obviously annihilates smooth functions, as it generates $H = U(1)$. Left operators are the Lie derivatives of the Killing directions on the sphere, and right operators are covariant derivatives, mapping between different spaces of sections of tensor bundles [16]. Notice that $\not{D}_L$ has the spin connection term, $+1$, displayed explicitly in (20), as opposed to $\not{D}_R$ where it is hidden in the definition of $\mathcal{R}$. The left version $\not{D}_L$ will act on $C^\infty(S^2, \mathbb{C})$ valued spinors, while $\not{D}_R$ will act on spinors valued on the ring of antiholomorphic forms, cf. section 4.

3. Equivalence of both approaches

That both expressions (20) of the Dirac operator are equivalent is not immediate but can be proven by use of the adjoint representation $Ad_G : \mathfrak{g} \to \mathfrak{g}$, defined as usual

$$Ad(g)X = gXg^{-1}, \quad \text{for} \quad X \in \mathfrak{g}. \tag{21}$$

If $Ad(g)_{ab}$ are the matrix elements in the basis $\{T_a\}$ we find readily the identity

$$Ad(g)_{ab}\mathcal{L}_b f(g) = -\mathcal{R}_a f(g), \quad \text{with } Ad_{ab}(g) = \frac{1}{2}\mathrm{Tr}(g\sigma_a g^{-1}\sigma_b) \tag{22}$$

notice that this is a point dependent frame rotation, indeed a local $SU(2)$ transformation. One may show that due to the embedding chosen (13) the columns of the adjoint matrix form an orthonormal frame $Ad_{ab}(g) = \xi^a_b$, being ξ^1, ξ^2 tangent to S^2 and $\xi^3 = x$ the normal vector. It follows that the induced metric on S^2 is the projector

$$P_{ab} = \delta_{ab} - x_a x_b = \xi^i_a \xi^i_b, \quad i = 1, 2. \tag{23}$$

The left Dirac operator written in the form (1) is $\not{D}_L = \sigma_a P_{ab}(\mathcal{L}_b + \frac{1}{2}\sigma_b)$, and observe that for functions $f \in C^\infty(S^2, \mathbb{C})$ the left operators are tangent $P_{ab}\mathcal{L}_b f = \mathcal{L}_a f$, a property which no longer holds for non-equivariant functions and that is central to the construction of the spinor space; it is easy to identify $\mathcal{L}_a$ as the usual angular momentum operators, thus $\mathcal{L}+\frac{1}{2}\sigma$ is the total angular momentum of the fermion. We now proceed to prove the aforementioned equivalence

$$g^{-1}\not{D}_L g = g^{-1}\sigma_a g P_{ab}\mathcal{L}_b = Ad(g)_{ca}\sigma_c P_{ab}\mathcal{L}_b = \xi^c_a\xi^i_a\sigma_c\xi^i_b\mathcal{L}_b = \sigma_i Ad(g)_{ib}\mathcal{L}_b = -\not{D}_R. \tag{24}$$

The equation above connects both versions and shows that they are unitarily equivalent through a local $SU(2)$ rotation of the spinor space, in particular it relates spinors $\Psi^{L,R}$

$$\Psi^L(g) = g\Psi^R(g) \tag{25}$$

4. Spinor space

Let $\mathcal{S}$ be the space of smooth spinor fields on S^2. The representation content of spinors is known for spin manifolds of the type G/H, at least in an abstract manner [10], and since the Dirac operator is a map $\not{D}:\mathcal{S}\to\mathcal{S}$ it is clear from (25) that in the left and right formulations spinors will have a different structure but the same representation content, for instance

$$\mathcal{S}^L = D^{\frac{1}{2}}\otimes C^\infty(S^2,\mathbb{C}) \in \bigoplus_{J=0}^{\infty} D^{J+\frac{1}{2}} \bigoplus_{J=1}^{\infty} D^{J-\frac{1}{2}}, \quad \mathcal{S}^L = \mathcal{S}^L_+\oplus\mathcal{S}^L_- \tag{26}$$

where each piece corresponds to a Weyl spinor. In the right action version spinors must be represented in a different manner, for definiteness let us we write

$$\Psi^{L,R} = \begin{pmatrix}\psi_+\\ \psi_-\end{pmatrix} \in \mathcal{S}^{L,R} \tag{27}$$

the distinction is that for $\mathcal{S}^R$ one has $\psi_+\in\Omega^0(S^2,\mathbb{C}) = C^\infty(S^2,\mathbb{C})$ and $\psi_-\otimes d\bar{\zeta}\in\Omega^{\bar{1}}(S^2,\mathbb{C})$ is an antiholomorphic 1-form in a complex local coordinate system $\{\zeta,\bar{\zeta}\}$. The components of right spinors Ψ^R are naturally identified with (possibly charged[2]) antiholomorphic differential forms [16], such identification is well known for Kähler manifolds [17], namely

$$\mathcal{S} = \bigoplus_{k=0}^{\dim_{\mathbb{C}}\mathcal{M}} \Omega^{\bar{k}}(\mathcal{M},\mathbb{C}). \tag{28}$$

Once we know the structure of S we may use Lichnerowicz's theorem [18]

$$\not{D}^2 = \Delta_{\mathcal{S}} + \frac{R}{4} \tag{29}$$

and the chirality[3] operator $\gamma = x\cdot\sigma$ to construct the nonzero spectrum of $\not{D}$ from the nonzero spectrum of $\not{D}^2$, the advantage being that $\not{D}^2$ is given entirely in terms of quadratic Casimir operators and hence its spectrum becomes straightforward to compute. Let $\Psi^{(\sigma)}_\Lambda$ be the eigenspinor of $\not{D}^2$ with eigenvalue Λ and let (σ) label degeneracy; $\Psi^{(\sigma)}_\Lambda$ has a definite chirality since $[\not{D}^2,\gamma]=0$ and we shall also verify that $\Lambda>0$. First of all we have

$$\not{D}^2\Psi^{(\sigma)}_\Lambda = \Lambda\Psi^{(\sigma)}_\Lambda, \tag{30}$$

[2] $U(1)$ charged forms describe spinors in the spin-c structures, these correspond physically to the motion of a fermion on a topological monople field [16].

[3] Which anticommutes with either Dirac operator $\{\not{D}^{L,R},\gamma\}=0$.

and under these conditions to each eigenspinor of $\not{D}^2$ correspond exactly two eigenspinors of $\not{D}$

$$\Psi_\Lambda^{(\sigma;\pm)} = \Psi^{(\sigma)} \pm \frac{\not{D}\Psi_\Lambda^{(\sigma)}}{\sqrt{\Lambda}}, \qquad \not{D}\Psi_\Lambda^{(\sigma;\pm)} = \pm\sqrt{\Lambda}\Psi_\Lambda^{(\sigma;\pm)} \tag{31}$$

Lichnerowicz's theorem implies that on a spin manifold with strictly positive curvature there cannot be zero modes, therefore on S^2 the full spectrum can be obtained from (31); this follows from (29) since a zero mode Ψ_0 would mean $\not{D}\Psi_0 = 0$, nevertheless by construction $\Delta_{\mathcal{S}}$ is a positive operator, hence $\langle\Psi_0, \Delta_{\mathcal{S}}\Psi_0\rangle \geq 0$, and since $R > 0$ equation (29) would lead to the contradiction $|\not{D}\Psi_0|^2 = \langle\Psi_0, \not{D}^2\Psi_0\rangle > 0$; from here it follows also $\Lambda > 0$. The spin Laplacian is $\Delta_{\mathcal{S}} = -g^{\mu\nu}\nabla_\mu\nabla_\nu$, which in our case of interest can be cast into the forms

$$\Delta_{\mathcal{S}}^L = \left(\mathcal{L}_a + \frac{P_{ab}\sigma_b}{2}\right)^2 = \left(\mathcal{L}_a + \frac{1}{2}\sigma_a\right)^2 - \left(\frac{\sigma_0}{2}\right)^2 \tag{32}$$

$$\Delta_{\mathcal{S}}^R = \mathcal{R}_i^2 = \mathcal{R}_a^2 - \mathcal{R}_0^2 \tag{33}$$

Clearly both (σ) and Λ are fully determined from the $SU(2), U(1)$ representation content of the spinor space, we remark that both $\mathcal{S}^R$, $\mathcal{S}^L$ have the same representation content. The full spectrum is symmetric around zero and can be found directly using $R = 2$, cf. appendix Appendix A, the degeneracy index (σ) counts the dimension of each irreducible representation appearing in (26)

$$\mathrm{Spec}(\not{D}) = \{\lambda \in \mathbb{Z}/\{0\}\}, \quad \sigma = 1, 2, \cdots, 2|\lambda| \tag{34}$$

5. The role of the constraint

Finally we should clarify what is the meaning of the differential operators we have introduced, for instance $\partial_1 = \frac{\partial}{\partial z_1}$ would usually imply that z_1 can be varied in the argument of a function $f(z_1, z_2, \bar{z}_1, \bar{z}_2)$ leaving all other arguments constant, but this is clearly impossible as we have the constraint (11). In a sense the constraint plays a trivial role since the left and right differential operators we are interested in are all tangent to S^3. A way to show that we may *formally* operate on $z_A, \bar{z}_B$ as if they were all independent is to consider a function f on S^3 as the restriction of a smooth extension $\tilde{f}$ to $\mathbb{C}^2$, in the spirit of [4]

$$\tilde{f}(z_A, \bar{z}_B) = f(z_A, \bar{z}_B)\Delta(z^\dagger z - 1) \in C^\infty(\mathbb{C}^2, \mathbb{C}), \quad \tilde{f} = f\big|_{S^3} \tag{35}$$

where $\Delta(y)$ is a smooth function that goes rapidly to zero[4] for $y \neq 0$ while $\Delta(0) = 1$ and f is formally extended considering $z_A, \bar{z}_B$ as independent variables. Now we may verify that the central property required is that the constraint $z^\dagger z - 1$ must be annihilated by the differential operators (understood as vector fields on $\mathbb{C}^2$), $\mathcal{R}_a(z^\dagger z - 1) = \mathcal{L}_a(z^\dagger z - 1) = 0$; it is very easy to check that this is indeed the case and we have for instance

$$\mathcal{R}_a[\tilde{f}]\big|_{S^3} = \mathcal{R}_a[f(z_A, \bar{z}_B)\Delta(z^\dagger z - 1)]\big|_{S^3} = \Delta(z^\dagger z - 1)\mathcal{R}_a[f(z_A, \bar{z}_B)]\big|_{S^3} = \mathcal{R}_a[f]. \tag{36}$$

Thus we may operate as if $z_A, \bar{z}_B$ were all independent variables and obtain the same result after restriction to S^3. As pointed out in [4] this is a consequence of the fact that $\mathcal{L}_a, \mathcal{R}_b$ may be understood as flows on S^3, and consider functions on S^3 as functions of its polar coordinates whereupon such differential operators depending only of these polar coordinates will act. This justifies the use of the computation rules $\frac{\partial z_A}{\partial z_B} = \frac{\partial \bar{z}_A}{\partial \bar{z}_B} = \delta_A^B$ and $\frac{\partial \bar{z}_A}{\partial z_B} = \frac{\partial z_A}{\partial \bar{z}_B} = 0$ throughout the calculations, notwithstanding the constraint.

[4] It is also possible to take the limit when Δ becomes singular, one such example is $k \to +\infty$ and $\Delta(y) = e^{-ky^2}$, the result (36) is independent of the Δ chosen.

Remarks

When a space has a high degree of symmetry and can be realised as a coset space there is a particularly simple treatment of field theory problems using representation theory. If the space admits a spin structure then it is feasible to construct the spinor space and spin bundle using representation theory and to solve the spectral problem for the Dirac operator essentially through quadratic Casimir operators of the involved groups, rendering the problem very simple. We have reviewed the details corresponding to S^2 and shown both approaches, based on left and right action differential operators, which are naturally induced from the group structure. One makes use of Lie derivatives and the other of covariant derivatives, in the first case the spin connection is displayed explicitly, and both are ultimately equivalent. Although we have discussed here only a simple example the general approach works for a wide variety of spaces G/H, and constitutes a significant computational advantage over the vielbein approach, in addition to the advantages provided by the natural use of the symmetry involved.

Acknowledgments

I. Huet acknowledges support under the SNI funding system, S. Gutiérrez is thankful to the UAM for continued financial support.

Appendix A. Curvature on Coset Spaces

In what follows we found it convenient to use the following notation: μ labels local coordinates on $\mathcal{M} = G/H$, a labels the basis of $\mathfrak{g}$ which splits into α, β that label $\mathfrak{h}$ and i, j, k, l, m that label the remaining elements of the basis (corresponding to tangent directions on G/H), notice that this choice implies $f_{i\alpha\beta} = 0$; we follow closely the discussion found in [18, 19]. Quite generally we may set up a local coordinate system x^μ on G/H and choose a representative on each coset class $g(x) \in G$ corresponding to the point labelled by x^μ, the associated right Maurer-Cartan form will be

$$e^a_\mu dx^\mu \otimes T_a = g^{-1}(x) dg(x) \in \Omega^1(G/H) \otimes \mathfrak{g}$$

such form decomposes accordingly into $e^a_\mu dx^\mu \otimes T_a = e^i_\mu dx^\mu \otimes T_i + e^\alpha_\mu dx^\mu \otimes T_\alpha = e^i \otimes T_i + e^\alpha \otimes T_\alpha$. The vielbein is the tangent part of the Maurer-Cartan form, that is $e^i = \theta^i_R$ is the vielbein for G/H, and taken together with the connection 1-form ${\omega^i}_k$ determine the curvature and torsion of $\mathcal{M}$ through Cartan's structure equations

$$de^i + {\omega^i}_k \wedge e^k = \mathfrak{T}^i \tag{A.1}$$
$$d{\omega^i}_k + {\omega^i}_j \wedge {\omega^j}_k = {\mathfrak{R}^i}_k \tag{A.2}$$

in these equations ${\omega^i}_k = \Gamma^i_{jk} e^j_\mu dx^\mu$ with Γ^i_{jk} the Christoffel symbols in the vielbein basis. The Levi-Civita connection is compatible with the G-invariant metric and has vanishing torsion $\mathfrak{T}^i = 0$. Equation (A.1) can be compared against the Maurer-Cartan equation for G

$$d\theta_L + \theta_L \wedge \theta_L = 0 \tag{A.3}$$

leading, after using Jacobi's identity, to the connection and curvature forms

$${\omega^i}_k = \frac{1}{2} {f^i}_{jk} e^j + {f^i}_{k\alpha} e^\alpha \tag{A.4}$$
$${\mathfrak{R}^i}_k = \frac{1}{4}(2{f^i}_{k\alpha} f^\alpha_{lm} + {f^i}_{kj} {f^j}_{lm} - {f^i}_{lj} {f^j}_{km}) e^l \wedge e^m \,. \tag{A.5}$$

IOP Conf. Series: Journal of Physics: Conf. Series **1010** (2018) 012010 doi:10.1088/1742-6596/1010/1/012010

If we denote by $Ad_{G/H}$ the restriction of the adjoint representation Ad_G to H within the tangent space of G/H [5], and the normalisation for the quadratic Casimir is fixed by $C_2 := T_a T_a$, then we find the Ricci scalar curvature to be

$$R = \frac{1}{2}\left(\mathrm{Tr}C_2(Ad_G) - \mathrm{Tr}C_2(Ad_H) - \mathrm{Tr}C_2(Ad_{G/H})\right). \tag{A.6}$$

In the particular case of symmetric spaces (where $f_{ijk} = 0$, meaning that the torsion of the canonically induced connection vanishes), such as S^n one has

$$R = f^i{}_{k\alpha} f^\alpha_{ik} = \mathrm{Tr}C_2(Ad_{G/H}) = \frac{1}{3}\left(\mathrm{Tr}C_2(Ad_G) - \mathrm{Tr}C_2(Ad_H)\right) \tag{A.7}$$

For S^2 we may use polar coordinates and choose for instance $g(\theta, \phi)$ be determined by

$$(z_1, z_2) = (\cos(\theta/2)e^{-i\frac{\phi}{2}}, \sin(\theta/2)e^{i\frac{\phi}{2}})$$

leading to $e^1 = d\theta$, $e^2 = \sin\theta d\phi$, $e^3 = -\cos\theta d\phi$ and $\omega^1{}_1 = \omega^2{}_2 = 0$, $\omega^2{}_1 = -\omega^1{}_2 = \cos\theta d\phi$. Either method produces $R = 2$ for the Ricci scalar curvature of the unit sphere, a standard result found e.g. in [13] and that was obtained also by use of group theory in (A.7) .

References

[1] Novoselov K S, Geim A K, Morosov S V, Jiang D, Katsnelson M I, Grigorieva I V, Dubonos S V and Firsov A A 2005 Two-dimensional gas of massless Dirac fermions in graphene *Nature* **438** 197–200

González J, Guinea F and Vozmediano M A H 1993 The electronic spectrum of fullerenes from the Dirac equation *Nucl. Phys.* B **406** 771–94

[2] Dolan B P 2003 The spectrum of the Dirac operator on coset spaces with homogeneous gauge fields *J. High Energy Phys.* JHEP05(2003)018

Dolan B P 2011 Duality in strongly interacting systems: $\mathcal{N} = 2$ SUSY Yang-Mills and the quantum Hall effect *Fortschr. Phys.* **59** 1174–86

[3] Abrikosov Jr A A 2002 Dirac operator on the Riemann sphere *Preprint* hep-th/0212134

Jayewardena C 1988 Schwinger model on S^2 *Helv. Phys. Acta* **61** 636–711

[4] Grosse H and Prešnajder P 1995 Dirac Operator on the Fuzzy Sphere *Lett. Math. Phys.* **33** 171–81

[5] Balachandran A P, Immirzi G, Lee J and Prešnajder P 2003 Dirac operators on coset spaces *J. Math. Phys.* **44** 4713–35 (*Preprint* hep-th/0210297)

[6] Camporesi R and Higuchi A 1996 On the eigenfunctions of the Dirac operator on spheres and real hyperbolic spaces *J. Geometry Phys.* **20** 1–18 (*Preprint* gr-qc/9505009)

[7] Carow-Watamura U and Watamura S 1997 Chirality and Dirac operator on noncommutative sphere *Commun. Math. Phys.* **183** 365–82 (*Preprint* hep-th/9605003)

[8] Bär C 1996 The Dirac operator on space forms of positive curvature *J. Math. Soc. Japan* **48** 69–83

[9] Trautman A 1995 The Dirac operator on hypersurfaces *Acta Phys. Pol.* B **26** 1283–310 (*Preprint* hep-th/9810018)

[10] Friedrich T 2000 *Dirac Operators in Riemannian Geometry* (Providence: Amer. Math. Soc.) chapters 3 and 4 pp 57–112

Huang J and Pandžić P 2006 *Dirac Operators in Representation Theory,* (Boston: Birkhäuser) chapter 3 pp 57–72

[11] Lawson Jr H B and Michelsohn M-L 1989 *Spin Geometry* (Princeton: Princeton University Press) chapter 2 pp 77–165

[12] Barut A O and Rączka R 1986 *Theory of Group Representations and Applications* (Singapore: World Scientific Publishing Co. Pte. Ltd.) chapter 7 pp 166–98

[13] Eguchi T, Gilkey P B and Hanson A J 1980 *Phys. Rep.* vol 66 ed E R Andrew *et al* (Amsterdam: North-Holland) chapters 2–5 pp 213–393

[14] O'Raifeartaigh L 1988 *Group Structure of Gauge Theories* (Cambridge: Cambridge University Press) chapters 1–3 pp 3–35

[15] Brink D M and Satchler G R 1968 *Angular Momentum* (London: Oxford University Press/Clarendon Press) chapter 2 pp 13–37

Varshalovich D A, Moskalev A N and Khersonskii V K 1989 *Quantum Theory of Angular Momentum* (Singapore: World Scientific Publishing Co. Pte. Ltd.) pp 72–80

[16] Dolan B P, Huet I, Murray S and O'Connor D 2008 A universal Dirac operator and noncommutative spin bundles over fuzzy complex projective spaces *J. High Energy Phys.* JHEP03(2008)029 (*Preprint* hep-th/0711.1347)
[17] Green M B, Schwarz J H and Witten E 1999 *Superstring Theory* vol 2 (Cambridge: Cambridge University Press) section 15.5 pp 444–52
[18] Lichnerowicz A 1963 Spineurs harmoniques *C. R. Acad. Sci. Paris* **257** pp 7–9
[19] Salam A and Strathdee J 1982 On Kaluza-Klein theory *Ann. of Phys.* **141** pp 316–52
Castellani L, Romans L J and Warner N P 1984 Symmetries of Coset Spaces and Kaluza-Klein Supergravity *Ann. of Phys.* **157** pp 394–407

IOP Conf. Series: Journal of Physics: Conf. Series **1010** (2018) 012011 doi:10.1088/1742-6596/1010/1/012011

P, C and T: Different Properties on the Kinematical Level

Valeriy V. Dvoeglazov

UAF, Universidad Autónoma de Zacatecas
Ap. Postal 636, Suc. 3, Zacatecas 98061, Zac., México

E-mail: valeri@fisica.uaz.edu.mx

Abstract. We study the discrete symmetries (P,C and T) on the kinematical level within the extended Poincaré Group. On the basis of the Silagadze research, we investigate the question of the definitions of the discrete symmetry operators both on the classical level, and in the secondary-quantization scheme. We study the physical contents within several bases: light-front formulation, helicity basis, angular momentum basis, and so on, on several practical examples. We analize problems in construction of the neutral particles in the the $(1/2,0)+(0,1/2)$ representation, the $(1,0)+(0,1)$ and the $(1/2,1/2)$ representations of the Lorentz Group. As well known, the photon has the quantum numbers 1^-, so the $(1,0)+(0,1)$ representation of the Lorentz group is relevant to its description. We have ambiguities in the definitions of the corresponding operators P, C; T, which lead to different physical consequences. It appears that the answers are connected with the helicity basis properties, and commutations/anticommutations of the corresponding operators, P, C, T, and C^2, P^2, $(CP)^2$ properties. This contribution is the review paper of my previous work [2, 3].

1. Introduction.

In his paper of 1992 Silagadze claimed: "It is shown that the usual situation when boson and its antiparticle have the same internal parity, while, fermion and its antiparticle have opposite parities, assumes a kind of locality of the theory. In general, when a quantum-mechanical parity operator is defined by means of the group extension technique, the reversed situation is also possible", Ref. [1]. Then, Ahluwalia *et al* proposed [4] the Bargmann-Wightman-Wigner-type quantum field theory, where, as they claimed, the boson and the antiboson have oposite intrinsic parities (see also [5]). Actually, this type of theories has been first proposed by Gelfand and Tsetlin (1956), Ref. [6]. In fact, it is based on the two-dimensional representation of the inversion group. They indicated applicability of this theory to the description of the system of K-mesons and the possible relations to the Lee-Yang paper. The (anti)comutativity of the discrete symmetry operations has also been investigated by Foldy and Nigam [7], who claimed that it is related to the eigenvalues of the charge operator. The relations of the Gelfand-Tsetlin construct to the representations of the anti-de Sitter $SO(3,2)$ group and the general relativity theory have also been discussed in subsequent papers of Sokolik. E. Wigner [8] presented some relevant results at the Istanbul School on Theoretical Physics in 1962. Later, Fushchich *et al* discussed the wave equations. Actually, the theory presented by Ahluwalia, Goldman and Johnson is the Dirac-like generalization of the Weinberg $2(2J+1)$-theory for the spin 1. The equations have also been presented in the Sankaranarayanan and Good paper of 1965, Ref. [9].

In [10] the theory in the $(\frac{1}{2},0)\oplus(0,\frac{1}{2})$ representation based on the chiral helicity 4-eigenspinors was proposed. The corresponding equations have been obtained, e. g., in [3]. However, later we found the papers by Ziino and Barut [11] and the Markov papers [12], which also have connections with the subject under consideration. The question of definitions of the discrete symmetries operators raised by Silagadze, has not yet been clarified in detail. Explicit examples are presented below and in the previous papers [2, 3, 13, 4, 5, 9, 10, 11, 12].

2. Helicity Basis and Parity.

The 4-spinors have been studied well when the basis has been chosen in such a way that they are eigenstates of the $\hat{\mathbf{S}}_3$ operator:

$$u_{\frac{1}{2},\frac{1}{2}} = N^+_{\frac{1}{2}}\begin{pmatrix}1\\0\\1\\0\end{pmatrix}, u_{\frac{1}{2},-\frac{1}{2}} = N^+_{-\frac{1}{2}}\begin{pmatrix}0\\1\\0\\1\end{pmatrix}, \tag{1}$$

$$v_{\frac{1}{2},\frac{1}{2}} = N^-_{\frac{1}{2}}\begin{pmatrix}1\\0\\-1\\0\end{pmatrix}, v_{\frac{1}{2},-\frac{1}{2}} = N^-_{-\frac{1}{2}}\begin{pmatrix}0\\1\\0\\-1\end{pmatrix}. \tag{2}$$

And, oppositely, the helicity basis case has not been studied almost at all (see, however, Refs. [14, 15]). Let me remind that the boosted 4-spinors $\Psi(\mathbf{p}) = column\,(\phi_R(\mathbf{p})\ \pm\phi_L(\mathbf{p}))$ in the common-used basis are the parity eigenstates with the eigenvalues of ± 1.

In the helicity spin basis the 2-eigenspinors of the helicity operator [16]

$$\frac{1}{2}\sigma\cdot\widehat{\mathbf{p}} = \frac{1}{2}\begin{pmatrix}\cos\theta & \sin\theta e^{-i\phi}\\ \sin\theta e^{+i\phi} & -\cos\theta\end{pmatrix}, \tag{3}$$

θ,ϕ are the angles of the spherical coordinate system, can be defined as follows [16, 17]:

$$\phi_{\frac{1}{2}\uparrow} \sim \begin{pmatrix}\cos\frac{\theta}{2}e^{-i\phi/2}\\ \sin\frac{\theta}{2}e^{+i\phi/2}\end{pmatrix},\quad \phi_{\frac{1}{2}\downarrow} \sim \begin{pmatrix}\sin\frac{\theta}{2}e^{-i\phi/2}\\ -\cos\frac{\theta}{2}e^{+i\phi/2}\end{pmatrix}, \tag{4}$$

for $h=\pm 1/2 =\uparrow\downarrow$ eigenvalues, respectively.

We start from the Klein-Gordon equation, generalized for describing the spin-1/2 particles (i. e., two degrees of freedom), $c=\hbar=1$:

$$(E+\sigma\cdot\mathbf{p})(E-\sigma\cdot\mathbf{p})\phi = m^2\phi\,. \tag{5}$$

It can be re-written in the form of the system of two first-order equations for 2-spinors as in the Sakurai book. At the same time, we observe that they may be chosen as the eigenstates of the helicity operator:

$$(E-(\sigma\cdot\mathbf{p}))\phi_\uparrow = (E-p)\phi_\uparrow = m\chi_\uparrow\,, \tag{6}$$
$$(E+(\sigma\cdot\mathbf{p}))\chi_\uparrow = (E+p)\chi_\uparrow = m\phi_\uparrow\,, \tag{7}$$

$$(E-(\sigma\cdot\mathbf{p}))\phi_\downarrow = (E+p)\phi_\downarrow = m\chi_\downarrow\,, \tag{8}$$
$$(E+(\sigma\cdot\mathbf{p}))\chi_\downarrow = (E-p)\chi_\downarrow = m\phi_\downarrow\,. \tag{9}$$

IOP Conf. Series: Journal of Physics: Conf. Series **1010** (2018) 012011 doi:10.1088/1742-6596/1010/1/012011

If the ϕ spinors are defined by the equation (4) then we can construct the corresponding $u-$ and $v-$ 4-spinors[1]

$$u_\uparrow = N_\uparrow^+ \begin{pmatrix} \phi_\uparrow \\ \frac{E-p}{m}\phi_\uparrow \end{pmatrix} = \frac{1}{\sqrt{2}} \begin{pmatrix} \sqrt{\frac{E+p}{m}}\phi_\uparrow \\ \sqrt{\frac{m}{E+p}}\phi_\uparrow \end{pmatrix}, \quad (10)$$

$$u_\downarrow = N_\downarrow^+ \begin{pmatrix} \phi_\downarrow \\ \frac{E+p}{m}\phi_\downarrow \end{pmatrix} = \frac{1}{\sqrt{2}} \begin{pmatrix} \sqrt{\frac{m}{E+p}}\phi_\downarrow \\ \sqrt{\frac{E+p}{m}}\phi_\downarrow \end{pmatrix}, \quad (11)$$

$$v_\uparrow = N_\uparrow^- \begin{pmatrix} \phi_\uparrow \\ -\frac{E-p}{m}\phi_\uparrow \end{pmatrix} = \frac{1}{\sqrt{2}} \begin{pmatrix} \sqrt{\frac{E+p}{m}}\phi_\uparrow \\ -\sqrt{\frac{m}{E+p}}\phi_\uparrow \end{pmatrix}, \quad (12)$$

$$v_\downarrow = N_\downarrow^- \begin{pmatrix} \phi_\downarrow \\ -\frac{E+p}{m}\phi_\downarrow \end{pmatrix} = \frac{1}{\sqrt{2}} \begin{pmatrix} \sqrt{\frac{m}{E+p}}\phi_\downarrow \\ -\sqrt{\frac{E+p}{m}}\phi_\downarrow \end{pmatrix}, \quad (13)$$

where the normalization to the unit (± 1) was used. One can prove that the matrix $P = \gamma^0 = \begin{pmatrix} 0 & I \\ I & 0 \end{pmatrix}$ can be used in the parity operator as that in the original Dirac basis. Indeed, the 4-spinors (10)-(13) satisfy the Dirac equation in the spinorial (Weyl) representation of the γ-matrices. Hence, the parity-transformed function $\Psi'(t, -\mathbf{x}) = P\Psi(t, \mathbf{x})$ must satisfy $[i\gamma^\mu \partial'_\mu - m]\Psi'(t, -\mathbf{x}) = 0$ with $\partial'_\mu = (\partial/\partial t, -\nabla_i)$. This is possible when $P^{-1}\gamma^0 P = \gamma^0$ and $P^{-1}\gamma^i P = -\gamma^i$. The P-matrix above satisfies these requirements, as in the textbook case [18].

Next, it is easy to prove that one can form the projection operators

$$\mathcal{P}_+ = +\sum_h u_h(\mathbf{p})\bar{u}_h(\mathbf{p}) = \frac{p_\mu \gamma^\mu + m}{2m}, \quad (14)$$

$$\mathcal{P}_- = -\sum_h v_h(\mathbf{p})\bar{v}_h(\mathbf{p}) = \frac{m - p_\mu \gamma^\mu}{2m}, \quad (15)$$

with the properties $P_+ + P_- = 1$ and $P_\pm^2 = P_\pm$. This permits us to expand the 4-spinors defined in the basis (1,2) in the linear superpositions of the helicity basis 4-spinors, and to find corresponding coefficients of the expansion:

$$u_\sigma(\mathbf{p}) = A_{\sigma h} u_h(\mathbf{p}) + B_{\sigma h} v_h(\mathbf{p}), \; v_\sigma(\mathbf{p}) = C_{\sigma h} u_h(\mathbf{p}) + D_{\sigma h} v_h(\mathbf{p}). \quad (16)$$

Multiplying the above equations by $\bar{u}_{h'}$, $\bar{v}_{h'}$ respectively, and using the normalization conditions, we obtain $A_{\sigma h} = D_{\sigma h} = \bar{u}_h u_\sigma$, $B_{\sigma h} = C_{\sigma h} = -\bar{v}_h u_\sigma$. Thus, the transformation matrix from the common-used basis to the helicity basis is

$$\begin{pmatrix} u_\sigma \\ v_\sigma \end{pmatrix} = \mathcal{U} \begin{pmatrix} u_h \\ v_h \end{pmatrix}, \quad \mathcal{U} = \begin{pmatrix} A & B \\ B & A \end{pmatrix} \quad (17)$$

Neither A nor B are unitary:

$$A = (a_{++} + a_{+-})(\sigma_\mu a^\mu) + (-a_{-+} + a_{--})(\sigma_\mu a^\mu)\sigma_3\,, \quad (18)$$

$$B = (-a_{++} + a_{+-})(\sigma_\mu a^\mu) + (a_{-+} + a_{--})(\sigma_\mu a^\mu)\sigma_3\,, \quad (19)$$

[1] One can also try to construct yet another theory differing from the ordinary Dirac theory. The 4-spinors may be *not* the eigenspinors of the helicity operator of the $(1/2, 0) \oplus (0, 1/2)$ representation space, cf. [10]. They might be the eigenstates of the *chiral* helicity operator introduced in [10].

IOP Conf. Series: Journal of Physics: Conf. Series **1010** (2018) 012011 doi:10.1088/1742-6596/1010/1/012011

where

$$a^0 = -i\cos(\theta/2)\sin(\phi/2) \in \Im m\,,\; a^1 = \sin(\theta/2)\cos(\phi/2) \in \Re e,$$
$$a^2 = \sin(\theta/2)\sin(\phi/2) \in \Re e\,,\; a^3 = \cos(\theta/2)\cos(\phi/2) \in \Re e, \quad (20)$$

and

$$a_{++} = \frac{\sqrt{(E+m)(E+p)}}{2\sqrt{2}m}\,,\; a_{+-} = \frac{\sqrt{(E+m)(E-p)}}{2\sqrt{2}m}\,, \quad (21)$$
$$a_{-+} = \frac{\sqrt{(E-m)(E+p)}}{2\sqrt{2}m}\,,\; a_{--} = \frac{\sqrt{(E-m)(E-p)}}{2\sqrt{2}m}\,. \quad (22)$$

However, $A^\dagger A + B^\dagger B = I$, so the matrix $\mathcal{U}$ is unitary. Please note that this matrix acts on the *spin* indices (σ, h), and not on the spinorial indices; it is the 4×4 matrix.

We now investigate the properties of the helicity-basis 4-spinors with respect to the discrete symmetry operations P, C and T. It is expected that $h \to -h$ under parity, as Berestetskiĭ, Lifshitz and Pitaevskiĭ claimed [19]. Indeed, if $\mathbf{x} \to -\mathbf{x}$, then the vector $\mathbf{p} \to -\mathbf{p}$, but the axial vector $\mathbf{S} \to \mathbf{S}$, that implies the above statement. The helicity 2-eigenspinors transform $\phi_{\uparrow\downarrow} \Rightarrow -i\phi_{\downarrow\uparrow}$ with respect to $\mathbf{p} \to -\mathbf{p}$, Ref. [17]. Hence,

$$Pu_\uparrow(-\mathbf{p}) = -iu_\downarrow(\mathbf{p})\,, Pv_\uparrow(-\mathbf{p}) = +iv_\downarrow(\mathbf{p})\,, \quad (23)$$
$$Pu_\downarrow(-\mathbf{p}) = -iu_\uparrow(\mathbf{p})\,, Pv_\downarrow(-\mathbf{p}) = +iv_\uparrow(\mathbf{p})\,. \quad (24)$$

Thus, on the level of classical fields, we observe that the helicity 4-spinors transform to the 4-spinors of the opposite helicity.

Also,

$$Cu_\uparrow(\mathbf{p}) = -v_\downarrow(\mathbf{p})\,, Cv_\uparrow(\mathbf{p}) = +u_\downarrow(\mathbf{p})\,, \quad (25)$$
$$Cu_\downarrow(\mathbf{p}) = +v_\uparrow(\mathbf{p})\,, Cv_\downarrow(\mathbf{p}) = -u_\uparrow(\mathbf{p})\,. \quad (26)$$

due to the properties of the Wigner operator $\Theta\phi^*_\uparrow = -\phi_\downarrow$ and $\Theta\phi^*_\downarrow = +\phi_\uparrow$, $\Theta_{[1/2]} = -i\sigma_2$. Similar conclusions can be obtained in the Fock space.

We define the field operator as follows:

$$\Psi(x^\mu) = \sum_h \int \frac{d^3\mathbf{p}}{(2\pi)^3}\frac{\sqrt{m}}{2E}[u_h(\mathbf{p})a_h(\mathbf{p})e^{-ip_\mu x^\mu} + v_h(\mathbf{p})b_h^\dagger(\mathbf{p})e^{+ip_\mu x^\mu}]\,. \quad (27)$$

The commutation relations are assumed to be the standard ones [21, 22, 18, 20] (compare with Refs. [3, 10]). If one defines $U_P\Psi(x^\mu)U_P^{-1} = \gamma^0\Psi(x^{\mu'})$, $U_C\Psi(x^\mu)U_C^{-1} = C\Psi^\dagger(x^\mu)$ and the anti-unitary operator of time reversal $(V_T\Psi(x^\mu)V_T^{-1})^\dagger = T\Psi^\dagger(x^{\mu''})$, then it is easy to obtain the corresponding transformations of the creation/annihilation operators:

$$U_P a_h(\mathbf{p})U_P^{-1} = -ia_{-h}(-\mathbf{p}),\;\; U_P b_h(\mathbf{p})U_P^{-1} = -ib_{-h}(-\mathbf{p}), \quad (28)$$
$$U_C a_h(\mathbf{p})U_C^{-1} = (-1)^{\frac{1}{2}+h}b_{-h}(\mathbf{p}),\;\; U_C b_h(\mathbf{p})U_C^{-1} = (-1)^{\frac{1}{2}-h}a_{-h}(-\mathbf{p}). \quad (29)$$

As a consequence, we obtain (provided that $U_P|0>=|0>$, $U_C|0>=|0>$)

$$U_P a_h^\dagger(\mathbf{p})|0>= U_P a_h^\dagger U_P^{-1}|0>= ia_{-h}^\dagger(-\mathbf{p})|0>= i|-\mathbf{p},-h>^+\,, \quad (30)$$
$$U_P b_h^\dagger(\mathbf{p})|0>= U_P b_h^\dagger U_P^{-1}|0>= ib_{-h}^\dagger(-\mathbf{p})|0>= i|-\mathbf{p},-h>^-\,, \quad (31)$$

and

$$U_C a_h^\dagger(\mathbf{p})|0>= U_C a_h^\dagger U_C^{-1}|0>= (-1)^{\frac{1}{2}+h} b_{-h}^\dagger(\mathbf{p})|0>= (-1)^{\frac{1}{2}+h}|\mathbf{p}, -h>^-, \tag{32}$$

$$U_C b_h^\dagger(\mathbf{p})|0>= U_C b_h^\dagger U_C^{-1}|0>= (-1)^{\frac{1}{2}-h} a_{-h}^\dagger(\mathbf{p})|0>= (-1)^{\frac{1}{2}-h}|\mathbf{p}, -h>^+ . \tag{33}$$

Finally, for the CP operation one should obtain:

$$U_P U_C a_h^\dagger(\mathbf{p})|0>= -U_C U_P a_h^\dagger(\mathbf{p})|0>= (-1)^{\frac{1}{2}+h} U_P b_{-h}^\dagger(\mathbf{p})|0>= = i(-1)^{\frac{1}{2}+h} b_h^\dagger(-\mathbf{p})|0>= i(-1)^{\frac{1}{2}+h}|-\mathbf{p}, h>^-, \tag{34}$$

$$U_P U_C b_h^\dagger(\mathbf{p})|0>= -U_C U_P b_h^\dagger(\mathbf{p}) = (-1)^{\frac{1}{2}-h} U_P a_{-h}^\dagger(\mathbf{p})|0>= = i(-1)^{\frac{1}{2}-h} a_h^\dagger(-\mathbf{p})|0>= i(-1)^{\frac{1}{2}-h}|-\mathbf{p}, h>^+ . \tag{35}$$

As in the classical case, the P and C operations anticommute in the $(\frac{1}{2},0)\oplus(0,\frac{1}{2})$ quantized case. This opposes to the theory based on the 4-spinor eigenstates of the chiral helicity (cf. [3, 7]). Since the V_T is an anti-unitary operator the problem must be solved after taking into account of the fact that the c-numbers should be put outside the hermitian conjugation *without* complex conjugation:

$$[V_T h A V_T^{-1}]^\dagger = [h^* V_T A V_T^{-1}]^\dagger = h[V_T A^\dagger V_T^{-1}]. \tag{36}$$

After applying this definition we obtain:[2]

$$V_T a_h^\dagger(\mathbf{p}) V_T^{-1} = +i(-1)^{\frac{1}{2}-h} a_h^\dagger(-\mathbf{p}), \tag{37}$$

$$V_T b_h(\mathbf{p}) V_T^{-1} = +i(-1)^{\frac{1}{2}-h} b_h(-\mathbf{p}). \tag{38}$$

Furthermore, we observed that the parity properties depend on the phase factor in the following definition:

$$U_P \Psi(t,\mathbf{x}) U_P^{-1} = e^{i\alpha}\gamma^0 \Psi(t,-\mathbf{x}). \tag{39}$$

Indeed,

$$U_P a_h(\mathbf{p}) U_P^{-1} = -ie^{i\alpha} a_{-h}(-\mathbf{p}), \tag{40}$$

$$U_P b_h^\dagger(\mathbf{p}) U_P^{-1} = +ie^{i\alpha} b_{-h}^\dagger(-\mathbf{p}). \tag{41}$$

From this, if $\alpha = \pi/2$ we obtain *opposite* parity properties of the creation/annihilation operators for particles and anti-particles:

$$U_P a_h(\mathbf{p}) U_P^{-1} = +a_{-h}(-\mathbf{p}), \tag{42}$$

$$U_P b_h(\mathbf{p}) U_P^{-1} = -b_{-h}(-\mathbf{p}). \tag{43}$$

However, the difference with the Dirac case still preserves (h transforms to $-h$). We find similar situation with the question of constructing the neutrino field operator (cf. with the Goldhaber-Kayser creation phase factor).

Next, we find the explicit form of the parity operator U_P and prove that it commutes with the Hamiltonian operator. We prefer to use the method described in [20, §10.2-10.3]. It is

[2] T should be chosen in such a way in order to fulfill $T^{-1}\gamma_0^T T = \gamma_0$, $T^{-1}\gamma_i^T T = \gamma_i$ and $T^T = -T$, as in Ref. [21].

based on the *anzatz* that $U_P = \exp[i\alpha\hat{A}]\exp[i\hat{B}]$ with $\hat{A} = \sum_s \int d^3\mathbf{p}[a^\dagger_{\mathbf{p},s}a_{-\mathbf{p}s} + b^\dagger_{\mathbf{p}s}b_{-\mathbf{p}s}]$ and $\hat{B} = \sum_s \int d^3\mathbf{p}[\beta a^\dagger_{\mathbf{p},s}a_{\mathbf{p}s} + \gamma b^\dagger_{\mathbf{p}s}b_{\mathbf{p}s}]$. On using the known operator identity

$$e^{\hat{A}}\hat{B}e^{-\hat{A}} = \hat{B} + [\hat{A},\hat{B}]_- + \frac{1}{2!}[\hat{A},[\hat{A},\hat{B}]] + \dots \tag{44}$$

and $[\hat{A},\hat{B}\hat{C}]_- = [\hat{A},\hat{B}]_+\hat{C} - \hat{B}[\hat{A},\hat{C}]_+$ one can fix the parameters α,β,γ such that one satisfies the physical requirements that a Dirac particle and its anti-particle have opposite intrinsic parities.

In our case, we need to satisfy the requirement that the operator should invert not only the sign of the momentum, but the sign of the helicity too. We may achieve this goal by the analogous postulate $U_P = e^{i\alpha\hat{A}}$ with

$$\hat{A} = \sum_h \int \frac{d^3\mathbf{p}}{2E}[a^\dagger_h(\mathbf{p})a_{-h}(-\mathbf{p}) + b^\dagger_h(\mathbf{p})b_{-h}(-\mathbf{p})]\,. \tag{45}$$

By direct verification, the requirement is satisfied provided that $\alpha = \pi/2$. You may compare this parity operator with that given in [18, 20] for Dirac fields:[3]

$$U_P = \exp\left[i\frac{\pi}{2}\int d^3\mathbf{p}\sum_s\Big(a(\mathbf{p},s)^\dagger a(\tilde{\mathbf{p}},s) + b(\mathbf{p},s)^\dagger b(\tilde{\mathbf{p}},s) - \right.$$
$$\left. -a(\mathbf{p},s)^\dagger a(\mathbf{p},s) + b(\mathbf{p},s)^\dagger b(\mathbf{p},s)\Big)\right]\,,\quad \text{(10.69) of [20]}. \tag{46}$$

By direct verification one can also come to the conclusion that our new U_P commutes with the Hamiltonian:

$$\mathcal{H} = \int d^3\mathbf{x}\Theta^{00} = \int d^3\mathbf{k}\sum_h[a^\dagger_h(\mathbf{k})a_h(\mathbf{k}) - b_h(\mathbf{k})b^\dagger_h(\mathbf{k})]\,, \tag{47}$$

i.e. $[U_P,\mathcal{H}]_- = 0$. Alternatively, we can try to choose other commutation relations [3, 10] for the set of bi-orthonormal states. The formulas of the theory have been presented in the $(\frac{1}{2},0)\oplus(0,\frac{1}{2})$ representation based on the chiral helicity 4-eigenspinors, see below. Next, the theory, which is based on a system of 6-component Weinberg-like equations in the $(1,0)\oplus(0,1)$ representation, has also been constructed. The results are similar. The papers by Ziino and Barut [11] and the Markov papers [12] have connections with the subject under consideration.

3. Chiral Helicity Construct and the Different Definition of the Charge Conjugate Operator on the Secondary Quantization Level.

In the chiral representation one can choose the spinorial basis (zero-momentum spinors) in the following way:

$$\lambda^S_\uparrow(\mathbf{0}) = \sqrt{\frac{m}{2}}\begin{pmatrix}0\\ i\\ 1\\ 0\end{pmatrix},\ \lambda^S_\downarrow(\mathbf{0}) = \sqrt{\frac{m}{2}}\begin{pmatrix}-i\\ 0\\ 0\\ 1\end{pmatrix}, \tag{48}$$

$$\lambda^A_\uparrow(\mathbf{0}) = \sqrt{\frac{m}{2}}\begin{pmatrix}0\\ -i\\ 1\\ 0\end{pmatrix},\ \lambda^A_\downarrow(\mathbf{0}) = \sqrt{\frac{m}{2}}\begin{pmatrix}i\\ 0\\ 0\\ 1\end{pmatrix}, \tag{49}$$

[3] Greiner used the following commutation relations $\left[a(\mathbf{p},s),a^\dagger(\mathbf{p}',s')\right]_+ = \left[b(\mathbf{p},s),b^\dagger(\mathbf{p}',s')\right]_+ = \delta^3(\mathbf{p}-\mathbf{p}')\delta_{ss'}$. One should also note that the Greiner form of the parity operator is not the unique one. Itzykson and Zuber [18] proposed another one differing by the phase factors from (10.69) of [20]. In order to find relations between those two forms of the parity operator one should apply additional rotation in the Fock space.

$$\rho^S_\uparrow(\mathbf{0}) = \sqrt{\frac{m}{2}}\begin{pmatrix}1\\0\\0\\-i\end{pmatrix}, \rho^S_\downarrow(\mathbf{0}) = \sqrt{\frac{m}{2}}\begin{pmatrix}0\\1\\i\\0\end{pmatrix}, \tag{50}$$

$$\rho^A_\uparrow(\mathbf{0}) = \sqrt{\frac{m}{2}}\begin{pmatrix}1\\0\\0\\i\end{pmatrix}, \rho^A_\downarrow(\mathbf{0}) = \sqrt{\frac{m}{2}}\begin{pmatrix}0\\1\\-i\\0\end{pmatrix}. \tag{51}$$

The indices $\uparrow\downarrow$ should be referred to the chiral helicity quantum number introduced in Ref. [10], $\eta = -h\gamma_5$. We use the notation of the previous papers on the subject. Ahluwalia and Grumiller used the helicity basis for the 2nd-type 4-spinors. The reader would immediately find the 4-spinors of the second kind $\lambda^{S,A}_{\uparrow\downarrow}(p^\mu)$ and $\rho^{S,A}_{\uparrow\downarrow}(p^\mu)$ in an arbitrary frame on using the boost operators:

$$\lambda^S_\uparrow(p^\mu) = \frac{1}{2\sqrt{E+m}}\begin{pmatrix}ip_l\\i(p^-+m)\\p^-+m\\-p_r\end{pmatrix}, \lambda^S_\downarrow(p^\mu) = \frac{1}{2\sqrt{E+m}}\begin{pmatrix}-i(p^++m)\\-ip_r\\-p_l\\(p^++m)\end{pmatrix}, \tag{52}$$

$$\lambda^A_\uparrow(p^\mu) = \frac{1}{2\sqrt{E+m}}\begin{pmatrix}-ip_l\\-i(p^-+m)\\(p^-+m)\\-p_r\end{pmatrix}, \lambda^A_\downarrow(p^\mu) = \frac{1}{2\sqrt{E+m}}\begin{pmatrix}i(p^++m)\\ip_r\\-p_l\\(p^++m)\end{pmatrix}, \tag{53}$$

and

$$\rho^S_\uparrow(p^\mu) = \frac{1}{2\sqrt{E+m}}\begin{pmatrix}p^++m\\p_r\\ip_l\\-i(p^++m)\end{pmatrix}, \rho^S_\downarrow(p^\mu) = \frac{1}{2\sqrt{E+m}}\begin{pmatrix}p_l\\(p^-+m)\\i(p^-+m)\\-ip_r\end{pmatrix}, \tag{54}$$

$$\rho^A_\uparrow(p^\mu) = \frac{1}{2\sqrt{E+m}}\begin{pmatrix}p^++m\\p_r\\-ip_l\\i(p^++m)\end{pmatrix}, \rho^A_\downarrow(p^\mu) = \frac{1}{2\sqrt{E+m}}\begin{pmatrix}p_l\\(p^-+m)\\-i(p^-+m)\\ip_r\end{pmatrix}. \tag{55}$$

Some of the 4-spinors are connected each other. The normalization of the spinors $\lambda^{S,A}_{\uparrow\downarrow}(p^\mu)$ and $\rho^{S,A}_{\uparrow\downarrow}(p^\mu)$ are the following ones:

$$\overline{\lambda}^S_\uparrow(p^\mu)\lambda^S_\downarrow(p^\mu) = -im \quad , \quad \overline{\lambda}^S_\downarrow(p^\mu)\lambda^S_\uparrow(p^\mu) = +im \quad , \tag{56}$$

$$\overline{\lambda}^A_\uparrow(p^\mu)\lambda^A_\downarrow(p^\mu) = +im \quad , \quad \overline{\lambda}^A_\downarrow(p^\mu)\lambda^A_\uparrow(p^\mu) = -im \quad , \tag{57}$$

$$\overline{\rho}^S_\uparrow(p^\mu)\rho^S_\downarrow(p^\mu) = +im \quad , \quad \overline{\rho}^S_\downarrow(p^\mu)\rho^S_\uparrow(p^\mu) = -im \quad , \tag{58}$$

$$\overline{\rho}^A_\uparrow(p^\mu)\rho^A_\downarrow(p^\mu) = -im \quad , \quad \overline{\rho}^A_\downarrow(p^\mu)\rho^A_\uparrow(p^\mu) = +im \quad . \tag{59}$$

All other conditions are equal to zero.

Implying that $\lambda^S(p^\mu)$ (and $\rho^A(p^\mu)$) answer for positive-frequency solutions; $\lambda^A(p^\mu)$ (and $\rho^S(p^\mu)$), for negative-frequency solutions, one can deduce the dynamical coordinate-space equations [3]:

$$i\gamma^\mu\partial_\mu\lambda^S(x) - m\rho^A(x) = 0\,, \tag{60}$$

$$i\gamma^\mu\partial_\mu\rho^A(x) - m\lambda^S(x) = 0\,, \tag{61}$$

$$i\gamma^\mu\partial_\mu\lambda^A(x) + m\rho^S(x) = 0\,, \tag{62}$$

$$i\gamma^\mu\partial_\mu\rho^S(x) + m\lambda^A(x) = 0\,. \tag{63}$$

They can be written in the 8-component form. This is just another representation of the Dirac-like equation in the appropriate Clifford Algebra. One can also re-write the equations into the two-component form, the Feynman-Gell-Mann equations. In the Fock space the operators of the charge conjugation and space inversions can be defined as in the previous Section. We imply the bi-orthonormal system of the anticommutation relations. As a result we have the following properties of the creation (annihilation) operators in the Fock space:

$$U^s_{[1/2]} a_\uparrow(\mathbf{p}) (U^s_{[1/2]})^{-1} = -i a_\downarrow(-\mathbf{p}), U^s_{[1/2]} a_\downarrow(\mathbf{p}) (U^s_{[1/2]})^{-1} = +i a_\uparrow(-\mathbf{p}), \tag{64}$$
$$U^s_{[1/2]} b^\dagger_\uparrow(\mathbf{p}) (U^s_{[1/2]})^{-1} = +i b^\dagger_\downarrow(-\mathbf{p}), U^s_{[1/2]} b^\dagger_\downarrow(\mathbf{p}) (U^s_{[1/2]})^{-1} = -i b_\uparrow(-\mathbf{p}), \tag{65}$$

that signifies that the states created by the operators $a^\dagger(\mathbf{p})$ and $b^\dagger(\mathbf{p})$ have different properties with respect to the space inversion operation, comparing with Dirac states (the case also regarded in [11]):

$$U^s_{[1/2]} |\mathbf{p}, \uparrow>^+ = +i| -\mathbf{p}, \downarrow>^+, U^s_{[1/2]} |\mathbf{p}, \uparrow>^- = +i| -\mathbf{p}, \downarrow>^-, \tag{66}$$
$$U^s_{[1/2]} |\mathbf{p}, \downarrow>^+ = -i| -\mathbf{p}, \uparrow>^+, U^s_{[1/2]} |\mathbf{p}, \downarrow>^- = -i| -\mathbf{p}, \uparrow>^- . \tag{67}$$

For the charge conjugation operation in the Fock space we have two physically different possibilities. The first one, *e.g.*,

$$U^c_{[1/2]} a_\uparrow(\mathbf{p}) (U^c_{[1/2]})^{-1} = +b_\uparrow(\mathbf{p}), U^c_{[1/2]} a_\downarrow(\mathbf{p}) (U^c_{[1/2]})^{-1} = +b_\downarrow(\mathbf{p}), \tag{68}$$
$$U^c_{[1/2]} b^\dagger_\uparrow(\mathbf{p}) (U^c_{[1/2]})^{-1} = -a^\dagger_\uparrow(\mathbf{p}), U^c_{[1/2]} b^\dagger_\downarrow(\mathbf{p}) (U^c_{[1/2]})^{-1} = -a^\dagger_\downarrow(\mathbf{p}), \tag{69}$$

in fact, has some similarities with the Dirac construct. The action of this operator on the physical states are

$$U^c_{[1/2]} |\mathbf{p}, \uparrow>^+ = +|\mathbf{p}, \uparrow>^-, U^c_{[1/2]} |\mathbf{p}, \downarrow>^+ = +|\mathbf{p}, \downarrow>^-, \tag{70}$$
$$U^c_{[1/2]} |\mathbf{p}, \uparrow>^- = -|\mathbf{p}, \uparrow>^+, U^c_{[1/2]} |\mathbf{p}, \downarrow>^- = -|\mathbf{p}, \downarrow>^+ . \tag{71}$$

But, one can also construct the charge conjugation operator in the Fock space which acts, *e.g.*, in the following way:

$$\widetilde{U}^c_{[1/2]} a_\uparrow(\mathbf{p}) (\widetilde{U}^c_{[1/2]})^{-1} = -b_\downarrow(\mathbf{p}), \widetilde{U}^c_{[1/2]} a_\downarrow(\mathbf{p}) (\widetilde{U}^c_{[1/2]})^{-1} = -b_\uparrow(\mathbf{p}), \tag{72}$$
$$\widetilde{U}^c_{[1/2]} b^\dagger_\uparrow(\mathbf{p}) (\widetilde{U}^c_{[1/2]})^{-1} = +a^\dagger_\downarrow(\mathbf{p}), \widetilde{U}^c_{[1/2]} b^\dagger_\downarrow(\mathbf{p}) (\widetilde{U}^c_{[1/2]})^{-1} = +a^\dagger_\uparrow(\mathbf{p}). \tag{73}$$

Therefore,

$$\widetilde{U}^c_{[1/2]} |\mathbf{p}, \uparrow>^+ = -|\mathbf{p}, \downarrow>^-, \widetilde{U}^c_{[1/2]} |\mathbf{p}, \downarrow>^+ = -|\mathbf{p}, \uparrow>^-, \tag{74}$$
$$\widetilde{U}^c_{[1/2]} |\mathbf{p}, \uparrow>^- = +|\mathbf{p}, \downarrow>^+, \widetilde{U}^c_{[1/2]} |\mathbf{p}, \downarrow>^- = +|\mathbf{p}, \uparrow>^+ . \tag{75}$$

Next, by straightforward verification one can convince ourselves about correctness of the the charge conjugation *commute* each other in the Fock space. For instance,

$$U^c_{[1/2]} U^s_{[1/2]} |\mathbf{p}, \uparrow>^+ = +i U^c_{[1/2]} | -\mathbf{p}, \downarrow>^+ = +i| -\mathbf{p}, \downarrow>^- \tag{76}$$
$$U^s_{[1/2]} U^c_{[1/2]} |\mathbf{p}, \uparrow>^+ = U^s_{[1/2]} |\mathbf{p}, \uparrow>^- = +i| -\mathbf{p}, \downarrow>^- . \tag{77}$$

The second choice of the charge conjugation operator answers for the case when the $\widetilde{U}^c_{[1/2]}$ and $U^s_{[1/2]}$ operations anticommute:

$$\widetilde{U}^c_{[1/2]} U^s_{[1/2]} |\mathbf{p}, \uparrow>^+ = +i \widetilde{U}^c_{[1/2]} | -\mathbf{p}, \downarrow>^+ = -i\,| -\mathbf{p}, \uparrow>^- \tag{78}$$
$$U^s_{[1/2]} \widetilde{U}^c_{[1/2]} |\mathbf{p}, \uparrow>^+ = -U^s_{[1/2]} |\mathbf{p}, \downarrow>^- = +i\,| -\mathbf{p}, \uparrow>^- . \tag{79}$$

Finally, the time reversal *anti-unitary* operator in the Fock space should be defined in such a way that the formalism to be compatible with the CPT theorem. If we wish the Dirac states to transform as $V(T)|\mathbf{p}, \pm 1/2 >= \pm | -\mathbf{p}, \mp 1/2 >$ we have to choose (within a phase factor), Refs. [18, 21]:

$$S(T) = \begin{pmatrix} \Theta_{[1/2]} & 0 \\ 0 & \Theta_{[1/2]} \end{pmatrix} . \tag{80}$$

Thus, in the first relevant case we obtain for the $\Psi(x^\mu)$ field, composed of $\lambda^{S,A}$ or $\rho^{A,S}$ 4-spinors

$$V^T a_\uparrow^\dagger(\mathbf{p})(V^T)^{-1} = a_\downarrow^\dagger(-\mathbf{p}),\ V^T a_\downarrow^\dagger(\mathbf{p})(V^T)^{-1} = -a_\uparrow^\dagger(-\mathbf{p}) \tag{81}$$

$$V^T b_\uparrow(\mathbf{p})(V^T)^{-1} = b_\downarrow(-\mathbf{p}),\ V^T b_\downarrow(\mathbf{p})(V^T)^{-1} = -b_\uparrow(-\mathbf{p}), \tag{82}$$

that is not surprising.

4. The Conclusions.

Thus, we proceeded as in the textbooks and defined the parity matrix as $P = \gamma_0$, because the helicity 4-spinors satisfy the Dirac equation, and the 2nd-type λ-spinors satisfy the coupled Dirac equations. Nevertheless, the properties of the corresponding spinors appear to be different with respect to the parity both on the first and the second quantization level. This result is compatible with the statement made by Berestetskii, Lifshitz and Pitaevskii. We defined another charge conjugation operator in the Fock space, which also transforms the positive-energy 4-spinors to the negative-energy ones. In this case the operators P and C commute (instead of the anticommutation in the Dirac case), that is related to the eigenvalues of the charge operator, as in the Foldy and Nigam paper.

References

[1] Z. K. Silagadze, *On the internal parity of anti-particles.* Sov. J. Nucl. Phys. **55** (1992) 392.
[2] V. V. Dvoeglazov, *Helicity basis and parity.* Int. J. Theor. Phys. **43** (2004) 1287.
[3] V. V. Dvoeglazov, *Different quantum field constructs in the* $(1/2, 0) \oplus (0, 1/2)$ *representation.* Mod. Phys. Lett. A**12** (1997) 2741.
[4] D. V. Ahluwalia, M. B. Johnson and T. Goldman, Phys. Lett. B**316** (1993) 102.
[5] V. V. Dvoeglazov, Int. J. Theor. Phys. **37** (1998) 1915, and references therein.
[6] I. M. Gel'fand and M. L. Tsetlin, Sov. Phys. JETP **4** (1957) 947; G. A. Sokolik, ibid. **6** (1958) 1170.
[7] B. P. Nigam and L. L. Foldy, Phys. Rev. **102** (1956) 1410.
[8] E. P. Wigner, in *Group Theoretical Concepts and Methods in Elementary Particle Physics – Lectures of the Istanbul Summer School of Theoretical Physics, 1962*, ed. F. Gürsey (Gordon and Breach, 1964).
[9] A. Sankaranarayanan and R. H. Good, jr., Nuovo Cim. **36**, (1965) 1303.
[10] N. D. S. Gupta, Nucl. Phys. B**4** (1967) 147; D. V. Ahluwalia, Int. J. Mod. Phys. A**11** (1996) 1855; V. Dvoeglazov, Hadronic J. **20** (1997) 435.
[11] G. Ziino, Ann. Fond. Broglie **14** (1989) 427; ibid **16** (1991) 343; A. Barut and G. Ziino, Mod. Phys. Lett. A**8** (1993) 1011; G. Ziino, Int. J. Mod. Phys. A**11** (1996) 2081.
[12] M. Markov, ZhETF **7** (1937) 579; ibid., p. 603; Nucl. Phys. **55** (1964) 130.
[13] D. V. Ahluwalia and M. Sawicki, Phys. Rev. D**47** (1993) 5161, nucl-th/9603019; Phys. Lett. B**335** (1994) 24, hep-th/9407095.
[14] Yu. V. Novozhilov, *Introduction to Elementary Particle Physics* (Pergamon, 1975) §4.3, 6.2.
[15] M. Jackob and G. C. Wick, Ann. Phys. **7** (1959) 404.
[16] D. A. Varshalovich, A. N. Moskalev and V. K. Khersonskiĭ, *Quantum Theory of Angular Momentum* (World Scientific, 1988), §6.2.5.
[17] V. V. Dvoeglazov, Fizika B**6** (1997) 111. x
[18] C. Itzykson and J.-B. Zuber, *Quantum Field Theory* (McGraw-Hill Book Co., 1980), Ch. 3-4.
[19] V. B. Berestetskiĭ, E. M. Lifshitz and L. P. Pitaevskiĭ, *Quantum Electrodynamics* (Pergamon Press, 1982), §16.
[20] W. Greiner, *Field Quantization* (Springer, 1996), Ch. 10.
[21] N. N. Bogoliubov and D. V. Shirkov, *Introduction to the Theory of Quantized Fields* (Wiley, 1980).
[22] S. Weinberg, *The Quantum Theory of Fields. Vol. I. Foundations.* (Cambridge UP, 1995).